Intelligent Manufacturing

This textbook introduces intelligent manufacturing concepts and strategies toward systems' best performance, especially concerning design and operations. It simplifies theory concepts through examples of real-life systems worldwide and presents many modern concepts such as sensor-based intelligent control systems, industrial decision-making, CAD/CAP/CAM integration in intelligent manufacturing, production control, augmented reality/mixed reality, cloud manufacturing, and manufacturing execution systems.

Intelligent Manufacturing: Concepts and Beyond covers the fundamental principles of intelligent manufacturing and provides a comprehensive resource for readers at different levels of expertise. This textbook is written in an infographic language, making complex concepts accessible to a broader audience. It includes many examples and numerical and practical exercises that reinforce learning and demonstrate how intelligent manufacturing systems work. It discusses contemporary technologies and trends, such as digital control and automation and machine/deep learning, and it provides a global perspective toward new emerging trends as it spans various technologies across different disciplines.

Written for both undergraduate and graduate courses, this textbook will take the reader from basic concepts and applications to advanced topics and can be the sole source to reach knowledge and explore future possibilities related to intelligent manufacturing techniques, methodologies, operations concepts, and beyond.

A solutions manual, PowerPoint slides, and figure slides are available for qualified textbook adoption.

Intelligent Manufacturing
Concepts and Beyond

Uzair Khaleeq uz Zaman, Atal Anil Kumar
and Aamer Ahmed Baqai

CRC Press
Taylor & Francis Group
Boca Raton London New York

CRC Press is an imprint of the
Taylor & Francis Group, an **informa** business

Designed cover image: Shutterstock

First edition published 2026
by CRC Press
2385 NW Executive Center Drive, Suite 320, Boca Raton FL 33431

and by CRC Press
4 Park Square, Milton Park, Abingdon, Oxon, OX14 4RN

CRC Press is an imprint of Taylor & Francis Group, LLC

ISBN: 978-1-032-77792-4 (hbk)
ISBN: 978-1-032-77873-0 (pbk)
ISBN: 978-1-003-48518-6 (ebk)

DOI: 10.1201/9781003485186

Typeset in Palatino
by codeMantra

Access the Instructor and Student Resources: www.routledge.com/9781032777924

Dedicated to the makers of tomorrow—may you design with intent, build with intelligence, and lead with purpose.

Contents

Foreword

In an age of rapid technological advancement, supply chain volatility, and the pursuit of net-zero targets, the manufacturing world finds itself at a crucial crossroads. The traditional paradigms of production are no longer sufficient. What is now demanded is agility, intelligence, and integration, qualities at the heart of *Intelligent Manufacturing: Concepts and Beyond*.

This textbook arrives at a timely juncture, offering a rich and accessible guide to the future of manufacturing. It captures the transition from conventional methods to digitally driven ecosystems, where cyber-physical systems, real-time analytics, and intelligent decision-making redefine what it means to design, plan, and produce. It uses global case studies and practical illustrations to explain complex ideas into digestible insights for students, educators, and professionals alike.

What makes this work particularly commendable is its panoramic vision. The authors skillfully interweave foundational principles with emerging technologies, ranging from CAD/CAM/CAPP integration to smart sensors, immersive reality, and cloud-based systems. Each chapter is not only theoretically rigorous but deeply grounded in practical relevance. Including modern pillars such as decision intelligence, machine learning, sustainability, and circular manufacturing illustrates a forward-looking mindset that aligns with the demands of Industry 4.0 and beyond.

This book is more than just a textbook. It is a toolkit for future-ready engineers and an enabler for organizations seeking to transform their manufacturing capabilities. It equips readers with the technical know-how and the strategic vision needed to thrive in a connected, data-driven industrial landscape.

For students, it will serve as a roadmap. For educators, it offers a structured and application-oriented curriculum. And for practitioners, it provides a wealth of ideas and strategies for implementation. In essence, this book empowers its readers not just to understand intelligent manufacturing, but to shape it.

Ali Siadat
Professor and Director
Laboratoire de Conception Fabrication Commande (LCFC)
Arts et Métiers Institute of Technology, Metz, France

Preface

In modern-day manufacturing, there are always going to be obstacles for manufacturers. These challenges include supply chain disruptions, shortage of skills, net-zero goals, and the continuing impacts of the coronavirus epidemic, to name a few. The most effective approach an enterprise can take is to foster resilience since the challenges manufacturers face are constantly increasing. This further entails not only raising productivity and quality but also being capable of adapting to rapidly changing demands and consumer expectations. Therefore, to ensure maximum performance, it is essential to utilize data and technology to create an optimized environment where intelligent manufacturing comes into play.

Moreover, Industry 4.0, the next generation of business, promises greater industrial flexibility, mass customization, better quality, and increased productivity. It enables businesses to meet the problems of producing products that are higher in quality, have shorter lead times to market, and are increasingly customized. Intelligent manufacturing is crucial in implementing Industry 4.0, wherein typical resources are transformed into intelligent objects, enabling them to see, respond, and behave in a smart environment.

The book *Intelligent Manufacturing: Concepts and Beyond* aims to introduce intelligent manufacturing concepts and strategies toward systems' best performance, especially concerning design and operations. This book simplifies theory concepts through examples of real-life systems worldwide. It presents many modern concepts such as industrial decision-making, intelligent control systems, CAD/CAPP/CAM integration in intelligent manufacturing, augmented reality/mixed reality, cloud manufacturing, and manufacturing execution systems. The topics cover the syllabus of undergraduate students of electrical, mechanical, mechatronics, and computer science, and the course is designed to be used as a text for the core and elective courses in manufacturing, automation, and controls. This textbook consists of 9 chapters and will serve as a comprehensive resource of education, inspiration, and empowerment for students and professionals in manufacturing industries. Furthermore, the descriptions are made in an infographic as much as possible to be absorbed by the reader more concisely and clearly.

Authors

Uzair Khaleeq uz Zaman, PhD, is an Associate Professor and Head of the Department of Mechanical Engineering at the National University of Sciences and Technology, Pakistan. He earned a PhD in industrial engineering at Arts et Métiers Institute of Technology, France, and has over 15 years of combined academic and industry experience. His core expertise includes smart manufacturing, industrial systems design, additive manufacturing, and AI-driven decision-making. Dr. Zaman has taught and conducted research at institutions including Georgia Tech-Lorraine (France), University of Alberta (Canada), Arts et Métiers ParisTech (France), Beihang University (China), ENSET Mohammedia (Morocco), and the University of Lorraine (France). He also focuses on integrating emerging technologies into engineering education, promoting sustainable practices, and leading curriculum reform. Passionate about outcome-based learning and industry collaboration, he mentors students and faculty to bridge research with real-world applications aligned with Industry 4.0.

Atal Anil Kumar, PhD, is a Professor at the University of Lorraine, Metz, France, where he is the Chair of Robotics for Sustainable and Agile Industry. He also did his post-doctoral research at the University of Luxembourg in the field of robotics. His research interests include nonlinear control, collaborative robots, robot kinematics and dynamics, and human–robot interaction. He has a bachelor's degree in mechatronics and an Erasmus master's in advanced robotics (with scholarship) from France and Italy. He has been awarded the research internship grant by the Indian Institute of Technology, Bombay (IIT-B) for performing research during his bachelor's degree. Dr. Atal has also worked as an R&D engineer in the automotive and steel manufacturing industries to gain significant hands-on experience and apply his theoretical knowledge.

Aamer Ahmed Baqai, PhD, is a Professor at the National University of Sciences and Technology, Pakistan. He earned a master's and a PhD in mechanical engineering at Arts et Métiers ParisTech, France. He has more than 26 years of experience in academia and industry. Dr. Baqai is a researcher of international repute and is a regular Research Visiting Professor at Arts et Métiers ParisTech, France. He has also acted as a foreign evaluator for PhD thesis students at Arts et Métiers ParisTech. His research interests include advanced manufacturing, manufacturing system design and optimization, reconfigurable manufacturing systems, process modeling and optimization, additive manufacturing, and decision-making in manufacturing.

Abbreviations

AHP Analytical hierarchy process
AI Artificial intelligence
AIDC Automatic identification and data capture
AM Additive manufacturing
ANN Artificial neural network
APP Aggregate production planning
AR Augmented reality
BOM Bill of materials
CACD Computer-aided conceptual design
CAD Computer-aided design
CAE Computer-aided engineering
CAPP Computer-aided process planning
CAM Computer-aided manufacturing
CFD Computational fluid dynamics
CIM Computer-integrated manufacturing
CMfg Cloud manufacturing
CNC Computer numerically controlled
CNN Convolutional neural network
CPS Cyber-physical system
CSR Corporate social responsibility
DCS Distributed control systems
DES Discrete event simulation
DMS Dedicated manufacturing system
DNC Direct numerical control
DI Decision intelligence
DL Deep learning
DSP Digital signal processor
DSS Decision support system
EOQ Economic order quantity
ERP Enterprise resource planning
FDC Factory data collection
FEA Finite element analysis
FEM Finite element modeling
FMC Flexible manufacturing cell
FMS Flexible manufacturing system
FoF Factory of the future
GA Genetic algorithm
GM Geometric modeling
GT Group technology
HVAC Heating, ventilation, and air-conditioning

IIoT	Industrial internet of things
IT	Information technology
IM	Intelligent manufacturing
IoT	Internet of Things
IIoT	Industrial internet of things
JIT	Just-in-time
LDR	Light-dependent resistor
LSTM	Long short-term memory
MAS	Multi-agent system
MCDM	Multi-criteria decision-making
MES	Manufacturing execution system
MIMO	Multiple-input multiple-output
ML	Machine learning
MPC	Model predictive control
MPS	Master production schedule
MR	Mixed reality
MRP	Material requirement planning
MRP-II	Manufacturing resource planning
MTS	Make-to-stock
NC	Numerical control
NCP	Numerical control programming
OEE	Overall equipment effectiveness
PdM	Predictive maintenance
PDM	Production data management
PID	Proportional-integral-derivative
PIR	Passive infrared
PLC	Programmable logic controller
PLM	Product lifecycle management
PO	Purchase order
PPC	Production planning and control
PR	Purchase requisition
RE	Reverse engineering
RFID	Radio frequency identification
RL	Reinforcement learning
ROP	Reorder point
RUL	Remaining useful life
RTD	Resistance temperature detector
SCADA	Supervisory control and data acquisition
SFC	Shop floor control
SDG	Sustainable development goal
SLAM	Simultaneous localization and mapping
SM	Smart manufacturing
SVM	Support vector machine
TIC	Total annual inventory cost
TOPSIS	Technique for order of preference by similarity to ideal solution

VM Virtual machine
VR Virtual reality
WC Work center
WIP Work in progress
WMS Warehouse management system
XAI Explainable AI
XR Extended reality
YOLO You only look once

Part 1

Concepts in Intelligent Manufacturing

1

Manufacturing Concepts and Intelligent Manufacturing

1.1 Introduction

In modern-day manufacturing, there are always going to be obstacles for manufacturers. These challenges include supply chain disruptions, shortage of skills, net-zero goals, and the continuing impacts of the coronavirus epidemic, to name a few. The most effective approach an enterprise can take is to foster resilience, since the challenges manufacturers face are constantly increasing. This further entails not only raising productivity and quality but also being capable of adapting to rapidly changing demands and consumer expectations. Therefore, to ensure maximum performance, it is essential to utilize data and technology to create an optimized environment where intelligent manufacturing (IM) comes into play. Moreover, with the advent of Industry 4.0, where automation and data are now essential to manufacturing, plants face increased pressure to implement IM.

But before understanding how IM operates, it is essential to comprehend what manufacturing is, the drivers for the global manufacturing industry outlook today, core manufacturing concepts such as the manufacturing life cycle and the associated differences between manufacturing process, manufacturing system, and manufacturing technology.

1.2 Manufacturing and the Global Manufacturing Industry Outlook

Manufacturing is a process of converting raw materials into applicable products. These days, the manufacturing industry is considered the pillar industry, not only being the determinant factor in economic competition but also being an essential symbol of economic/financial independence. The eventual goal for any manufacturing enterprise is to realize the aim of TQCSEK (i.e., faster Time-to-market, higher Quality, lower Cost, better Service, clean Environment, and high Knowledge) by employing three core strategies in their manufacturing paradigms, i.e., attaining "digital muscle," obtaining

DOI: 10.1201/9781003485186-2

"digital momentum," and expressing genuine commitment to improving the world by becoming "social supers." Today's manufacturing industry outlook is governed by the strategies mentioned above, which are explained below.

1.2.1 Digital Muscle

The current business climate is driving manufacturing companies to build digital muscle to attain increased flexibility in global supply chains. Applying tools such as artificial intelligence (AI), cloud computing, advanced analytics, robotics, and additive manufacturing (AM) to the value chain can increase visibility and transparency, allowing manufacturers to make faster changes to operations and respond quickly to market-based threats or opportunities.

1.2.2 Digital Momentum

These days, manufacturers are turning to partnerships to attain digital momentum. Since nothing can be achieved in isolation, mobilizing partnerships within their ecosystem can drive targeted business goals, from bolstering a traditional area (e.g., improving customer experience) to adding new capabilities (e.g., creating new business models). The manufacturers who attain digital momentum are also referred to as "forward-leaning manufacturers" as they have the capability to include digital technologies in their existing manufacturing processes and are also able to create "customer stickiness."

1.2.3 Social Super

The term social super originates from the concept of corporate social responsibility (CSR), which means that manufacturers are ready to develop or change their products/services to impact the environment positively. Transitioning to renewable energy sources is one example of being a social super, where the commitment is toward clean and green energy.

The United Nations in 2015 launched the 17 Sustainable Development Goals (SDGs) and 169 targets to transform the world by 2030.[1] The three core strategies can be attained by employing SDGs such as SDG 7: Affordable and clean energy, SDG 9: Industry, innovation and infrastructure, SDG 12: Responsible consumption and production, and SDG 17: Partnerships for the goals, to name a few.

1.3 Manufacturing Lifecycle

The manufacturing life cycle is analogous to the product life cycle and is characterized by four distinct stages: product introduction, growth, maturity, and decline. A generic manufacturing life cycle follows the concept of "cradle to grave,"

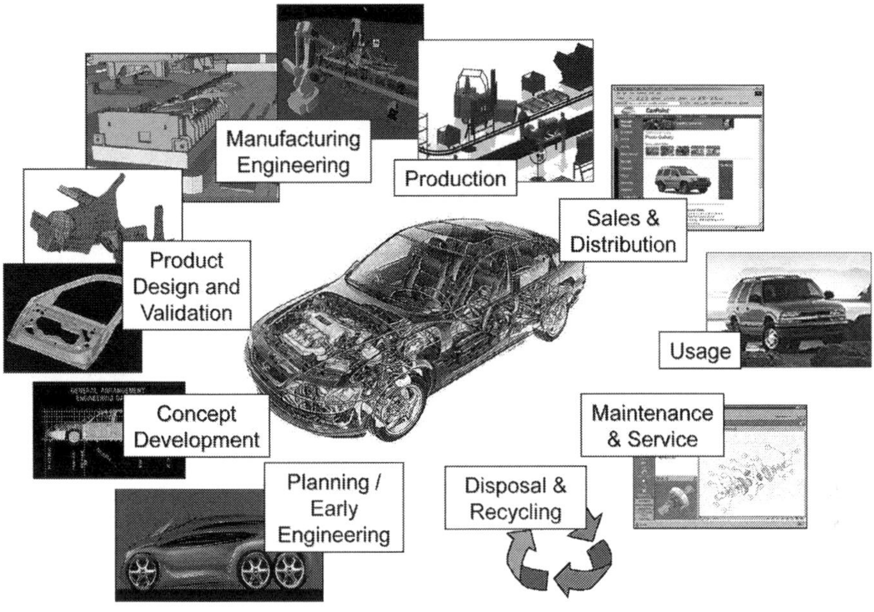

FIGURE 1.1
Manufacturing life cycle.

which aims to consider the entire life cycle of a product, from the raw materials used in its production to its disposal or recycling, as shown in Figure 1.1.

1.4 Basic Manufacturing Concepts

1.4.1 Manufacturing as an Economic and Technical Process[2]

It is possible to view manufacturing as both an economic and a technical process. For the former, manufacturing can be defined as the process of converting raw materials through one or more processing and/or assembly steps into products with a higher value. However, the latter uses physical and/or chemical processes to modify a starting material's geometry, characteristics, and/or appearance to create parts or finished goods. For instance, in a simple stamping process, an automated or manual stamping press punches holes in a sheet metal (technical process). In contrast, the same sheet with holes (of greater value) can be used in a cylinder head assembly (economic process).

1.4.2 Manufacturing System

A manufacturing system is composed of manufacturing objects, manufacturing resources, and workers used in the manufacturing process. It can also

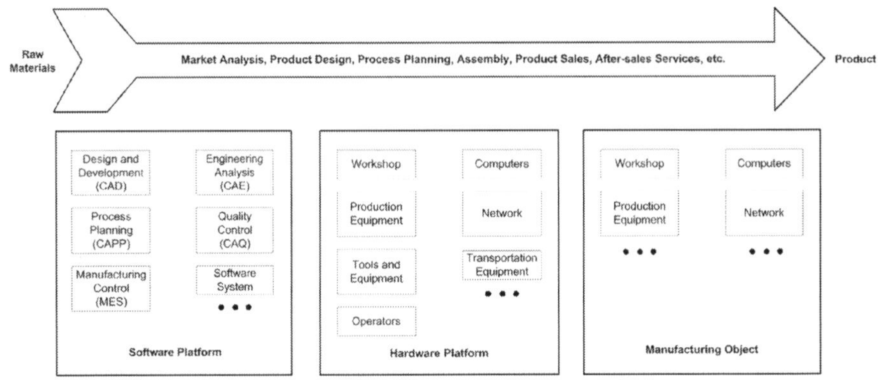

FIGURE 1.2
Schematic diagram of a manufacturing system.

be seen as a collection of machines, tools, and processes to produce goods or services. It typically includes the steps of raw materials acquisition, processing, assembly, and distribution. A schematic of the manufacturing system is shown in Figure 1.2.

1.4.3 Manufacturing Technology

Manufacturing technology lays the foundation for creating innovative products with high quality and competitive cost advantages, enabling efficient service delivery to the market. It plays a key role in supporting the manufacturing industry by enhancing productivity, precision, and flexibility.

Modern manufacturing technology integrates various disciplines, including mechanical and structural technology, design and process technology, computer control and assistive technology, automation technology, and information technology. This integration fosters seamless operations across different stages of production.

Furthermore, manufacturing technology applies these advanced techniques to multiple areas, such as market development, product supply and marketing, design, processing, testing, production management, and after-sales service. This holistic approach ensures that products meet high standards while optimizing efficiency and responsiveness to market demands.

1.5 Automation in Manufacturing Systems

International Business Machines (IBM) Corporation defines automation as the application of technology, programs, robotics, or processes to achieve an outcome with minimal human input.[3] This implies that automation includes creating such systems or processes that execute preset activities in response to particular triggers, situations, or algorithms. In short, it means delegating jobs to machines.

The history of automation dates to the ancient times when humans used tools such as pulleys, levers, and hand tools, to simplify their tasks. From water wheels in the 1st century to windmills and water mills in the 9th century, humans continued to simplify their tasks. However, it was not until the 17th century that industrial automation took a significant turning point with the advent of inventions such as steam engines and internal combustion engines. Sectors like transportation, healthcare, and the economy grew exponentially because of the mass production of materials such as plastics, cotton, glass, paper, and metals. Moving on into the era of 1900 to the 1950s, the high demand for instruments and controls was responded to by electrification and the use of industrial controllers. Equipment such as electrical substations, solid-state digital logic modules, hardwired programmable logic controllers (PLCs), high-pressure boilers, and electric motor-powered steam turbines increased the output of manufacturing facilities by more than 30%. Furthermore, with the invention of microprocessors in 1971, computer-associated hardware prices dropped considerably, opening doors for rapid growth of digital controls in the manufacturing industries. Advancements in computer technology continue to this day, with robotic automation systems augmenting almost every existing assembly and manufacturing process. In a nutshell, today's manufacturing technology landscape can be divided into two broad automation types: fixed and flexible. Table 1.1 shows the two broad automation types based on the kind of production and operational characteristics.

The fixed automation includes continuous operation and mass production flow, with the sequence of operations fixed by the equipment configuration (which implies very low product variety). Such a system is also known as a dedicated manufacturing system (DMS) whose advancements over time can be understood via a typical example of the Ford auto production line, which had the capability of producing 9,000 units of its "Model A" in a day in 1929,[4] to current production capabilities, where 59.5 vehicles per minute were produced by Chinese manufacturers alone in 2024.[5] Other fixed automation examples include manufacturing standard parts, bearings, tools, etc.

Flexible automation, on the other hand, addresses the challenges such as demand diversification and offers production capability of medium variety with high precision, but in small to medium-sized batches [also referred to as the flexible manufacturing system (FMS)]. In such an automation, no time is lost for changeovers from one design to the next, as re-programming the system and changing the physical setup (tooling, fixtures, machine settings) doesn't result in any lost production time. Moreover, rather than needing to be manufactured in batches, the system can also generate different combinations and schedules of parts or products.

TABLE 1.1

Broad Types of Industrial Automation and Operational Characteristics

Automation Type	Production Type	Operational Characteristics
Fixed automation	Continuous flow operation	• Flow process from beginning to end • Sensor is used for monitoring or control • Application of the integrated control and optimization strategy • Full realization of the computerized operation
	Mass production	• Combination of machine tools and automatic machines • Automatic transmission devices • Partial or fully automatic assembly • Robot is used for activities like welding, assembling, and spray paint • Includes process monitoring or control of the computer
Flexible automation	Batch production	• Numerical Control (NC) • Computer Numerical Control (CNC) • Direct Numerical Control (DNC) • Adaptive control • The robot application • Flexible Manufacturing Cell or System (FMC/FMS) • Computer-Integrated Manufacturing System (CIMS)
	Single production	• NC • CNC

Due to numeric control (NC) being part of flexible automation, FMS is considered an advanced version of a sometimes quoted third automation type, i.e., programmable automation. Since a program is used to control the operation sequence of a machine, such automation produces in small batches and can handle high product variety. Figure 1.3 shows the two broad automation types in terms of product volume and product variety.

1.6 Industry 4.0 and Intelligent Manufacturing[6]

Building upon the foundation of automation in manufacturing systems discussed in Section 1.5, the evolution of automation has been closely intertwined with major technological leaps known as industrial revolutions. Each revolution marked a paradigm shift in how goods were produced, influencing automation strategies significantly. Industry 1.0, the first paradigm shift, transitioned society from agriculture to an industrial focus through

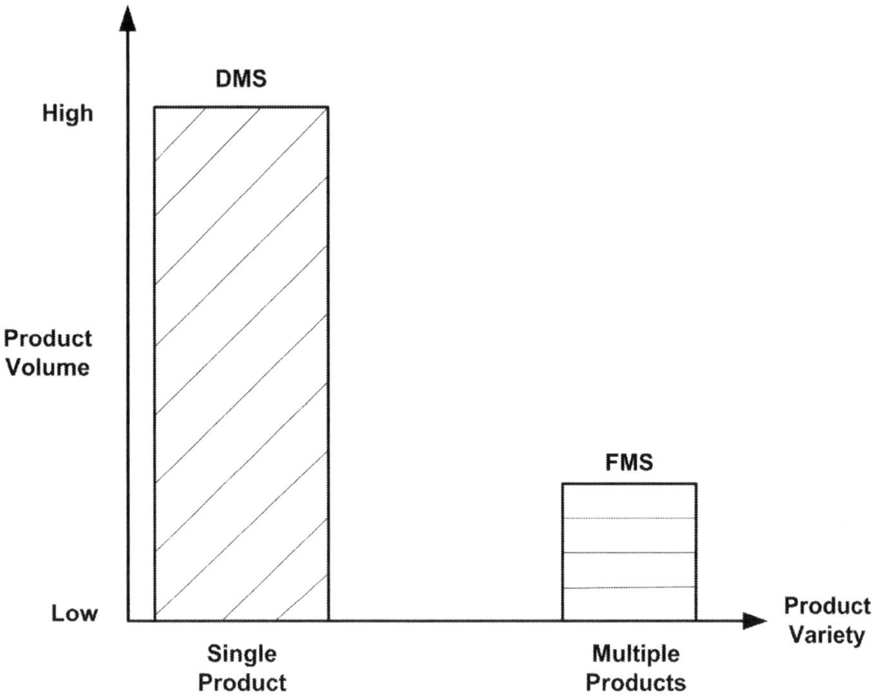

FIGURE 1.3
Fixed and flexible automation in terms of product variety and product volume.

water- and steam-powered machines. The primary goal was maximizing product volume with minimal product variety, laying the groundwork for fixed automation systems. Industry 2.0 introduced electricity and mechanical devices, shifting demand toward both variety and volume, which influenced early developments in flexible automation. Industry 3.0 marked the rise of digital technologies, reducing product lifecycles and adding delivery time as a critical dimension of manufacturing efficiency. This era saw the widespread adoption of PLCs and computer numerical control (CNC) systems, enhancing automation flexibility.

Currently, Industry 4.0 represents the integration of cyber-physical systems (CPS), driven by disruptive technologies like the Internet of Things (IoT), cloud computing, big data analytics, and AI. These advancements enable real-time data sharing, intelligent decision-making, and smart manufacturing systems that enhance quality, customization, and time-to-market, further evolving fixed and flexible automation paradigms. Figure 1.4 shows the four stages of the Industrial Revolution.

Since introducing intelligence into manufacturing has become an important topic for industries and researchers in the last few decades, terminologies such as smart manufacturing (SM) and IM in the context of Industry 4.0

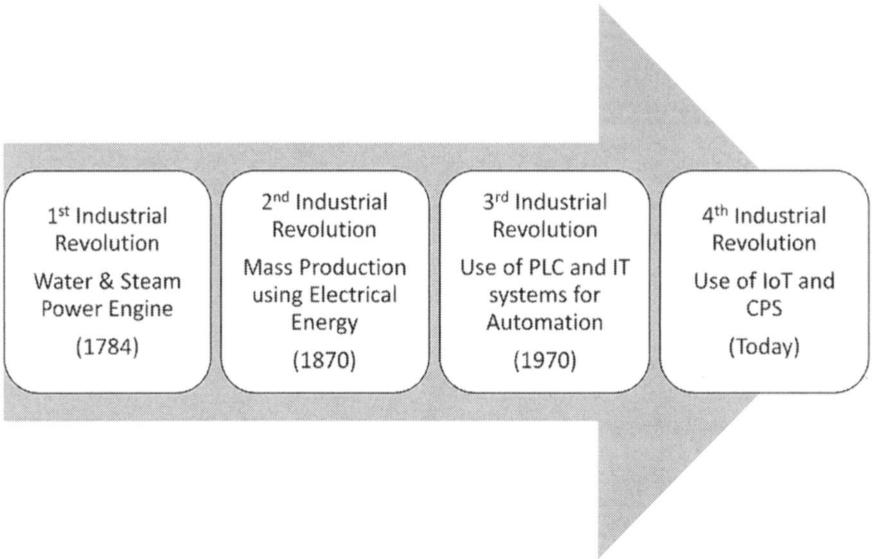

FIGURE 1.4
Four industrial revolutions.

are widely used. SM/IM is a broad concept of manufacturing that optimizes production and product transactions by fully using advanced information and manufacturing technologies. The complete lifecycle of a product and the associated value chain is covered by IM, starting from the conceptual design to production, delivery, and eventually recycling/disposal. IM also integrates real-time customer/user input/feedback, making the manufacturing process adaptive, agile, and intelligent. The interconnection of every step in the manufacturing process is, hence, the core objective associated with IM. Figure 1.5 shows the framework of Industry 4.0 IM systems.

1.6.1 Factory of the Future (FoF)

Since the early 20th century, researchers and practitioners have been idealizing what the factory of the future (FoF) will look like. Questions such as how future production environments will be and how machines and humans will communicate with each other are repeatedly asked. Today, products have shorter lifecycles, markets are dynamic, safety and reliability concerns are increasing, lifecycles are becoming more concise, and customers' awareness regarding quality is at its peak. In such scenarios, modern manufacturing systems, environments, and processes are affected directly and indirectly. Moreover, knowing how humans will come to terms with such an interaction is imperative with the complex interaction of machines, processes, and components. Nevertheless, such a scenario requires optimized human–machine

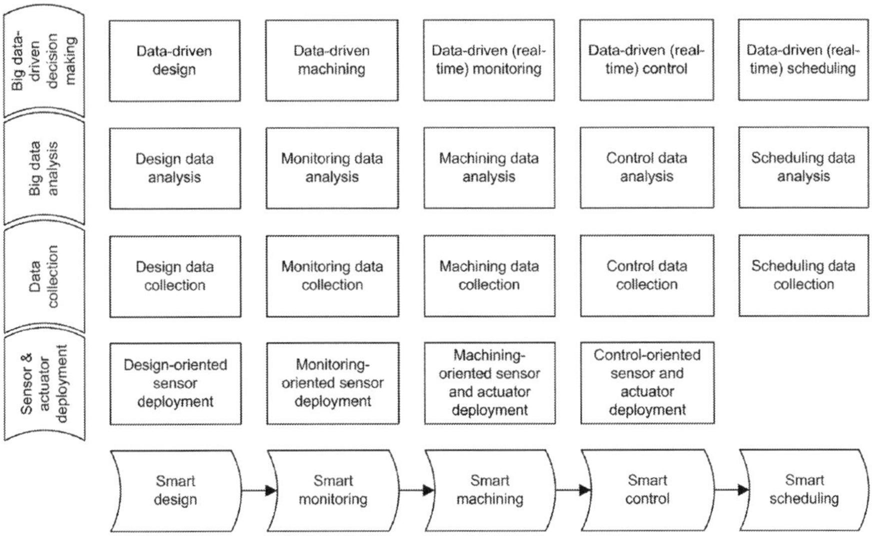

FIGURE 1.5
Conceptual framework of Industry 4.0 smart manufacturing systems.

operations and the interconnection of the "digital" and "real" worlds to get a bird's eye view of the complete value chain for increased production with greater efficiency using fewer resources.

FoF is, therefore, the true application of the paradigms of SM/IM wherein the production facilities and manufacturing plants are striving toward systems that are fully connected but also adaptive, autonomous, dependable, intelligent, scalable, collaborative, agile, and sustainable. Furthermore, the IM traits in the paragraphs mentioned above, along with digitization, bring forth new decision support systems (DSS) that utilize real-time production data and assist in revamping the entire product lifecycle. Concepts such as open value chain, flexible production, human-centered manufacturing, innovative business models (such as crowdsourcing, anything-as-a-service, and symbiotic ecosystems), and local initiatives [from Smart Manufacturing Leadership Coalition (SMLC), Industrial Internet Consortium (IIC), Japan's e-factory, German Industrie 4.0, intelligent manufacturing (China)] have become the backbone of the FoF.

1.6.2 Key Pillars of Industry 4.0 and Intelligent Manufacturing

Technological advancements common to IM are viewed as a new generation of information technology (IT) comprising Industry 4.0, IoT, big data, cloud and fog computing, AI, machine learning (ML), digital twins, and CPS. The nine pillars of Industry 4.0 that are making the manufacturing paradigm smarter and more intelligent are listed in Table 1.2.

TABLE 1.2

Nine Pillars of Industry 4.0

S. No.	Type of Pillar	Description
1	Big data and analytics	Big data and analytics enable the collection and comprehensive evaluation of data from many different sources to support real-time decision-making, optimization of production quality, energy saving, and improvement in equipment service
2	Autonomous robots	An autonomous robot is used to perform autonomous production methods more precisely and also work in places where human workers are restricted to work. Autonomous robots can complete given tasks precisely and intelligently within the given time limit, and also focus on safety, flexibility, and versatility
3	Simulation	Simulations will be used more extensively in plant operations to leverage real-time data to mirror the physical world in a virtual model, which can include machines, products, and humans, thereby reducing machine setup times and increasing quality
4	System integration: Horizontal and vertical system integration	The paradigm of Industry 4.0 is essentially outlined by three dimensions of integration: (a) horizontal integration across the entire value creation network, (b) vertical integration and networked manufacturing systems, and (c) end-to-end engineering across the entire product life cycle
5	Industrial IoT (IIoT)	The IoT is also known as the Internet of Everything (IoE), which consists of the Internet of Service (IoS), Internet of Manufacturing Services (IoMs), Internet of People (IoP), an embedded system, and Integration of Information and Communication Technology (IICT). Collectively, it is called IIoT
6	Cybersecurity and CPS	Reliable communications and sophisticated identity and access management of machines and users are important for Industry 4.0 to address the issue of cybersecurity threats, which increases dramatically with the increased connectivity and use of standard communication protocols
7	Cloud	Cloud-based IT platform serves as a technical backbone for connecting and communicating manifold elements of the Application Centre Industry 4.0. "Digital production" is a concept of connecting different devices to the same cloud to share information with one another. It can be extended to a set of machines from the shop floor as well as the entire plant
8	Additive manufacturing	Additive manufacturing methods will be widely used to produce small batches of customized products that offer construction advantages, such as complex, lightweight designs. High-performance, decentralized additive manufacturing systems will reduce transport distances and stock on hand
9	Augmented reality	Augmented reality–based systems can support various services, such as selecting parts in a warehouse and sending repair instructions over mobile devices

1.6.3 Future Perspectives of Intelligent Manufacturing in Industry 4.0

For a future Industry 4.0 IM system, the perspectives can be many. However, to simplify, the future perspectives can be categorized into "smart machines" (like smart robots and CPS-enabled smart machine tools), "smart design" [such as augmenting design with virtual reality (VR)/augmented reality (AR), and coupling computer-aided design (CAD)/computer-aided manufacturing (CAM) software with smart physical prototype systems], "smart monitoring" via sensors, "smart control" using cyber-physical production-control systems, "smart scheduling" through big data analytics, "cloud-based solutions" [using cloud computing and service-oriented architecture (SOA)], and "agent-based implementations" utilizing multi-agent technologies.

1.7 Highlights

Highlights of the chapter on intelligent manufacturing and Industry 4.0 are as follows:

a. Introduction to Modern Manufacturing Challenges: Manufacturers face dynamic challenges such as supply chain disruptions, skill shortages, net-zero goals, and post-pandemic impacts. The key to resilience is adaptability through IM, driven by data and technology, especially with the rise of Industry 4.0.
b. Global Manufacturing Outlook: The focus is on achieving TQCSEK: faster time-to-market, higher quality, lower costs, better service, clean environment, and strong knowledge management through:
   - Digital Muscle: Leveraging AI, robotics, cloud computing, and additive manufacturing for flexibility.
   - Digital Momentum: Building partnerships for digital transformation.
   - Social Super: Emphasizing sustainability and corporate social responsibility.
c. Manufacturing Life Cycle: Covers the full "cradle-to-grave" process, from raw materials to product disposal, with stages of introduction, growth, maturity, and decline.
d. Core Manufacturing Concepts:
   - Economic vs. Technical Processes: Transforming raw materials into high-value products.
   - Manufacturing Systems: Integration of machines, resources, and human input.

- Manufacturing Technology: Combines automation, IT, and design technologies for efficiency.

e. Automation in Manufacturing Systems: It is defined as reducing human intervention using robotics, PLCs, and digital controls.
   - Fixed Automation: High-volume, low-variety production (e.g., Ford Model A assembly lines).
   - Flexible Automation: Adapts to changing demands with minimal downtime, supported by CNC and programmable systems.

f. Industry 4.0 Evolution: It traces industrial revolutions from Industry 1.0 (steam power) to Industry 4.0, characterized by CPS, IoT, big data, and AI for smart, data-driven manufacturing.

g. Intelligent Manufacturing (IM): IM optimizes production through real-time data integration, covering the entire product lifecycle, from design to recycling. It emphasizes agility, customer feedback, and adaptive processes.

h. Factory of the Future (FoF): It envisions highly connected, autonomous, and sustainable factories with optimized human–machine collaboration, leveraging smart systems for efficiency and resource optimization.

i. Key Pillars of Industry 4.0: Includes AI, ML, IoT, cloud computing, digital twins, CPS, and big data analytics, which form the backbone of IM.

j. Future Perspectives of IM:
   - Smart Machines: CPS-enabled robots and tools.
   - Smart Design: Integrating VR/AR with CAD/CAM.
   - Smart Monitoring and Control: Using sensors, big data, and real-time analytics.
   - Cloud-Based Solutions and Agent-Based Systems: Enhancing flexibility and automation.

Conclusively, this chapter establishes the critical role of IM and Industry 4.0 in shaping the future of manufacturing, focusing on sustainability, efficiency, and technological integration.

Exercises

P1.1 What are the key challenges faced by modern-day manufacturers, and how can intelligent manufacturing (IM) help overcome them?

P1.2 Define the concept of TQCSEK and explain its significance in modern manufacturing.

P1.3 What are the four stages of the manufacturing life cycle, and how do they relate to product sustainability?

P1.4 Compare Fixed Automation and Flexible Automation in terms of product variety and production volume.

P1.5 Describe the four industrial revolutions and their impact on manufacturing automation.

P1.6 **Application-based problem**

A company produces 10,000 identical smartphone cases daily using fixed automation. They plan to switch to flexible automation to introduce three new case designs. How will this shift impact production processes, and what advantages might the company gain?

Notes

1 United Nations. (n.d.). *Transforming Our World: The 2030 Agenda for Sustainable Development.* Retrieved February 7, 2025, from https://sdgs.un.org/2030agenda

2 Groover, M. P. (2015). *Automation, Production Systems, and Computer-Integrated Manufacturing* (4th ed.). New Jersey: Pearson.

3 IBM. (n.d.). *What Is Automation?* Retrieved February 7, 2025, from https://www.ibm.com/think/topics/automation

4 Ford Motor Company. (n.d.). *The 1928 Ford Model A.* Retrieved February 8, 2025, from https://corporate.ford.com/articles/history/the-1928-ford-model-a.html

5 China Association of Automobile Manufacturers (CAAM). (n.d.). *Statistics.* Retrieved February 8, 2025, from http://en.caam.org.cn/Index/lists/catid/69.html

6 Zaman, U. K. U., Siadat, A., Baqai, A. A., Naveed, K., & Kumar, A. (Eds.). (2023). Introduction to Industry 4.0 and smart manufacturing. In *Handbook of Manufacturing Systems and Design: An Industry 4.0 Perspective* (Chapter 1), (pp. 3–8). Taylor & Francis, ISBN: 9781003327523.

2

Computer-Aided Design (CAD), Computer-Aided Process Planning (CAPP), Computer-Aided Manufacturing (CAM), and Intelligent Manufacturing

2.1 Introduction

Modern manufacturing ecosystems demand agility, precision, and intelligence. From initial conceptual sketches to complex multi-axis machining, the design-to-manufacturing pipeline must be tightly integrated. The trio of CAD (Computer-Aided Design), CAPP (Computer-Aided Process Planning), and CAM (Computer-Aided Manufacturing) plays a pivotal role in this journey, providing a digital backbone that supports everything from geometric modeling to intelligent process automation. Complete integration is shown in Figure 2.1, representing computer-integrated engineering.

This chapter explores these domains individually and collectively, discussing their evolution, roles, integration mechanisms, and advancement toward intelligent manufacturing paradigms powered by data standards like STEP and AI-driven decision systems.

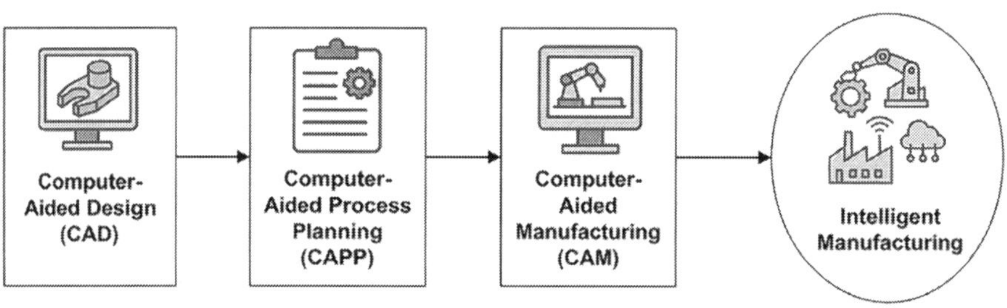

FIGURE 2.1
Computer-integrated engineering.

 DOI: 10.1201/9781003485186-3

2.2 Computer-Aided Design (CAD)

CAD refers to the use of computer systems to assist in creating, modifying, analyzing, and optimizing a design. Since the 1980s, CAD has been extensively utilized to create 2D engineering schematics and 3D models of physical components. Beyond visual representation, CAD is crucial in defining engineering workflows from conceptualization to final product design. It performs both static and dynamic analyses of assemblies to improve manufacturability.

The origin of CAD dates to the 1960s, when designers began using computers for 3D construction and NC programming, particularly in the aerospace and automotive sectors. Due to the high cost of computing at that time, large industries predominantly adopted early CAD systems. During the 1970s, CAD transitioned from traditional 2D drawings to more advanced solid modeling by the 1980s. Notable early modeling tools included Romulus by ShapeData, Uni-Solid from Unigraphics, and CATIA, a surface modeler by Dassault Systems. Autodesk introduced Autocad in 1982, marking a major milestone with a 2D drafting system, followed by the introduction of feature-based modeling in Pro/Engineer in 1988.

The late 1980s and early 1990s witnessed further advancements with the emergence of robust 3D modeling kernels such as B-rep, ACIS by Spatial Technology Inc., and Parasolid by Shape Data. Popular CAD tools like SolidWorks (1995), Solid Edge (1996), and IronCAD (1998) followed. Entering the 21st century, CAD systems began integrating functional modules to support broader engineering applications.

2.2.1 Functions of CAD

The functions of CAD can be categorized into computer-aided conceptual design (CACD), computer-aided geometric modeling (GM), computer-aided engineering (CAE), and computer-aided drafting.

2.2.1.1 Computer-Aided Conceptual Design (CACD)

CACD is used to create orthographic projections, schematics, layouts, and annotations. This function is commonly used in mechanical part drawings. CACD supports the early design phase by helping engineers explore design alternatives using functional decomposition, promoting systematic innovation in domains like automotive and aerospace. Figure 2.2 gives an example of CACD.

Example: In designing an electric vehicle's cooling system, CACD helps decompose the high-level function "regulate temperature" into sub-functions like "circulate coolant" and "monitor temperature," enabling structured design.

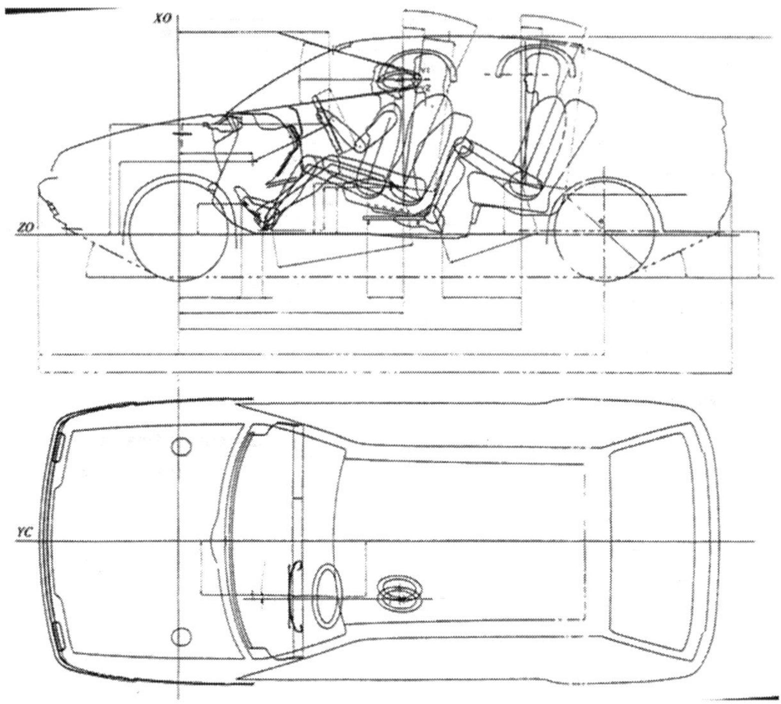

FIGURE 2.2
Computer-aided conceptual design (CACD).

2.2.1.2 Computer-Aided Geometric Modeling

GM refers to a set of techniques focused on creating a precise mathematical description of an object's geometry. It serves as the foundation for numerous engineering and manufacturing applications in the CAE, including mass property calculations, mechanism analysis, finite element modeling (FEM), and NC programming. The mathematical models in GM vary as per the application at hand, and each application is represented based on different geometric orientations, with a few requiring simple entities such as lines and arcs and others requiring curves and surfaces. The significant GM systems are as follows:

i. Wireframe Modeling: These represent objects using only lines and curves to illustrate edges and contours, making them suitable for conceptual visualization without emphasis on detailed physical properties. It is the least complex method in 3D representation and is used for defining complex solid objects.
ii. Surface Modeling: These define the outer surfaces of an object without representing its internal volume. These are more visually

realistic than wireframes and are commonly used in applications like automotive body design and styling.

iii. Solid Modeling: It represents the complete 3D volume of an object, including physical properties such as mass and material. It supports mathematical operations and is the most widely used in engineering contexts, particularly for mechanical part design, simulation, and integration with CAM systems.

Parametric and feature-based modeling is used in surface modeling, which is a powerful approach in 3D CAD that allows designers to create models defined by parameters such as length, diameter, angle, and other dimensional values. These parameters can be easily modified, enabling quick and efficient design updates without rebuilding the model from scratch. Feature-based modeling complements this by constructing geometry through a series of intuitive operations or "features" such as extrude, revolve, fillet, and chamfer. Each feature contributes to the part's final shape and can be edited independently, offering a highly flexible and systematic method for developing complex designs. This modeling technique supports design intent and promotes reusability and consistency across similar components.

Figure 2.3 shows the different types of GM systems.

Example 2.1: Parametric and Feature-Based Modeling of a Mechanical Bracket with Mounting Holes

A mechanical bracket can be efficiently designed in parametric and feature-based modeling by defining key parameters and constructing geometry through sequential features. For instance, consider a flat rectangular bracket with four mounting holes. The bracket's base is first created by sketching a 2D rectangle and using an 'extrude' feature to give it thickness. Dimensional parameters such as length (e.g., 150 mm), width (100 mm), and thickness (10 mm) are assigned and stored, allowing easy modification later. The mounting holes are then added using 'cut-extrude' operations, defined by parameters such as hole diameter and distances from the edges (e.g., 10 mm from each side). Each geometric feature is independently editable, including fillets on the corners or chamfers around the holes. Suppose a design change is needed, such as increasing the bracket's length or adjusting hole spacing. In that case, only the associated parameters are updated, and the model automatically regenerates to reflect the new values.

The notable CAD softwares include Autocad, SolidWorks, CATIA, PTC Creo, and Autodesk Fusion 360. A summary of CAD models with corresponding example use cases is given in Table 2.1.

One of the primary advantages of CAD software is improved accuracy, as CAD software reduces manual drafting errors through precise digital tools and geometric constraints, thus leading to more reliable and consistent designs.

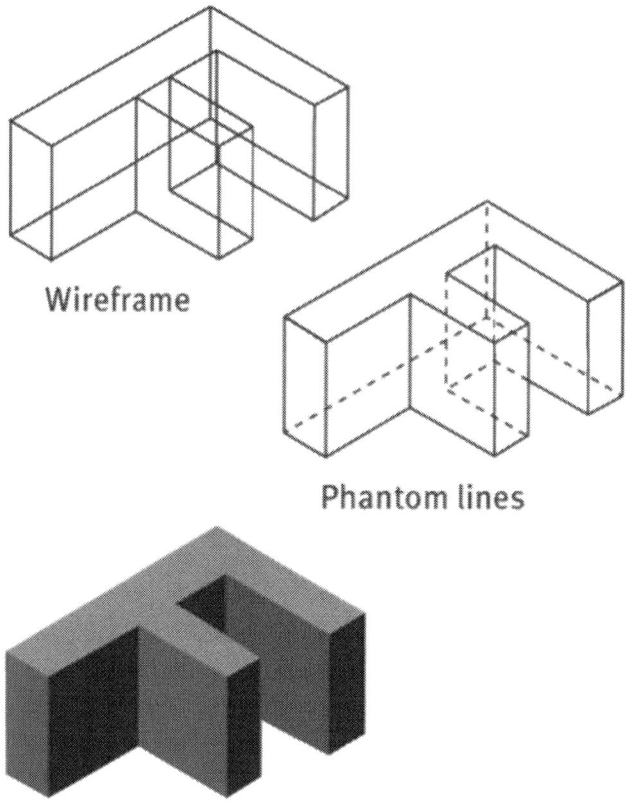

FIGURE 2.3
Different GM systems.

TABLE 2.1
Types of CAD Models

Model Type	Description	Example Use Case
2D CAD	Flat drawings with dimensions	Floor plans, circuit diagrams
3D wireframe	Lines and curves representing edges	Conceptual sketches, frame structures
Surface modeling	Surfaces without solid volume	Car bodies, consumer electronics
Solid modeling	Volume-based representation	Mechanical parts, tooling, assemblies

It also enhances productivity, as CAD allows rapid modifications, efficient documentation, and quick iteration during the design process. CAD offers better visualization by enabling realistic rendering and simulations, allowing designers to evaluate form, fit, and function before committing to physical prototypes. Additionally, data reusability is facilitated by standardized design libraries, templates, and modular components, promoting efficiency and consistency across projects. Finally, CAD systems are designed for seamless integration with CAM and CAPP, supporting automated manufacturing and process planning, streamlining the transition from design to production.

2.2.1.3 Computer-Aided Engineering (CAE)

CAE refers to the use of computer software to support engineering analysis tasks. CAE tools assist in simulating, validating, and optimizing product designs created in CAD, allowing engineers to predict a product's behavior in real-world conditions without the need for physical prototypes. CAE is a critical element of the product development lifecycle, enabling virtual testing, design verification, and decision support through advanced numerical methods such as Finite Element Analysis (FEA), Computational Fluid Dynamics (CFD), and Multibody Dynamics (MBD). For example, a heat exchanger designed in CAD is analyzed in a CAE environment to simulate temperature distribution and ensure structural integrity under thermal loads.

The workflow in CAE follows a step-by-step process. First, the CAD models (from GM stage) are transferred into the CAE environment, often with feature simplification. 'Meshing' is then performed, in which the model geometry is discretized into finite elements (mesh) suitable for numerical analysis. The 'Material and Boundary Condition Setup' is next done, where the designers assign material properties, loads, constraints, and environmental conditions. The CAE software then solves the mathematical equations governing the system behavior, followed by visualization of results (e.g., stress, strain, temperature, and velocity) through plots and animations. Finally, design feedback is also used, through which insights are fed back to GM and CACD stages or further down to the CAPP (Section 2.3) stage to refine design and process plans.

CAE tools often involve high computational demands, particularly when dealing with complex, nonlinear, or coupled problems such as thermal–structural interactions or fluid-structure dynamics. Moreover, the interpretation of simulation results requires skilled expertise. Without proper understanding, there is a risk of misleading or incorrect drawings and conclusions.

Example: An automotive engineer uses ANSYS to simulate the crashworthiness of a car bumper. After importing the 3D model from CAD software, the structure is meshed and subjected to impact loading conditions. The analysis reveals high stress in a specific region, prompting a design revision to increase thickness. This validated design is then passed to CAM for CNC toolpath generation.

2.2.1.4 Computer-Aided Drafting

CAD tools also provide detailed 2D engineering drawings derived from 3D models. These drawings are essential for documentation, inspection, and quality control. For instance, a gearbox housing modeled in 3D is automatically converted into detailed orthographic projections with dimensions, surface finishes, and Geometric, Dimensioning, and Tolerancing (GD&T) annotations.

2.3 Computer-Aided Process Planning (CAPP)

Computer-Aided Process Planning (CAPP) is the use of computer systems to aid in the planning of manufacturing processes for a given part or product. It serves as a critical bridge between product design (CAD) and manufacturing execution (CAM), translating engineering designs into detailed process plans and manufacturing instructions.

CAPP systems automate the decision-making in selecting processes, machining parameters, tools, fixtures, and work sequences, aiming to optimize production efficiency, quality, and cost. The primary objectives of CAPP are to streamline and standardize the creation of manufacturing process plans. By automating this critical function, CAPP aims to significantly reduce planning time and minimize human errors, leading to more consistent and efficient production workflows. Another key objective is to ensure the optimal utilization of machines, tools, and materials, contributing to cost reduction and improved productivity. CAPP systems also enhance a manufacturing system's adaptability to changes, such as design revisions or shifts in production volume. Furthermore, effective CAPP solutions support seamless integration with CAD, CAM, and Enterprise Resource Planning (ERP) systems, enabling a more connected and responsive manufacturing environment.

2.3.1 Types of CAPP Systems

The types of CAPP systems are listed in Table 2.2.

TABLE 2.2

Types of CAPP Systems

Type	Description
Variant CAPP	Based on Group Technology (GT), retrieves and modifies standard process plans for similar part families
Generative CAPP	Automatically generates process plans using decision logic, rules, and manufacturing databases
Semi-generative CAPP	Combines human input with automated planning capabilities
Dynamic CAPP	Adapts real-time plans based on resource availability, production conditions, or feedback

2.3.2 CAPP Workflow

CAPP workflow is given as follows:

a. Input from CAD: The CAD model provides geometric and material data of the part.
b. Part Classification (in variant systems): The part is identified and matched to a predefined family.
c. Process Selection: Decisions are made regarding the manufacturing method (e.g., turning, milling, casting).
d. Operation Sequencing: A logical order of operations is determined based on geometry and features.
e. Tool and Machine Selection: Suitable tools and machines are selected based on the required operations.
f. Parameter Assignment: Speeds, feeds, depths of cut, and tolerances are set based on material and operation.
g. Process Plan Generation: The system outputs a detailed plan that includes work instructions, setup sheets, and routing information.

Example 2.2

A **Generative CAPP** system receives a 3D model of a prismatic aluminum bracket. It identifies features such as holes and slots, selects drilling and milling operations, assigns appropriate tooling, and generates a complete process sheet including machine setup, tools needed, cutting parameters, etc. (Table 2.3).

The Generative CAPP for a Prismatic Aluminum Bracket is as shown below:

a. Input and Feature Recognition:
   Input: 3D CAD model of the bracket.
b. Material: Aluminum (e.g., 6061-T6).
c. Geometry Features Identified:

TABLE 2.3

Operation Sequence for Prismatic Aluminum Bracket

Operation	Description	Tools	Parameters
1	Rough face milling	Face mill	High feed rate, 2mm depth
2	Finish face milling	End mill	Low feed rate, 0.2mm depth
3	Pocket milling (if applicable)	Flat end mill	Adaptive toolpath
4	Hole drilling	Twist drill Ø10mm	Peck drilling cycle
5	Chamfering	Chamfer tool	1mm chamfer depth
6	Deburring	Manual or edge tool	N/A

   a. Flat base and side faces (prismatic shape).
   b. Through holes for mounting (e.g., Ø10 mm).
   c. Pockets or cutouts (if applicable).
   d. Chamfers or fillets.

d. Process Planning Strategy
e. Machining Type: Subtractive (CNC milling+drilling).
f. Setup Orientation:
   a. Primary Setup: Flat base down, accessible top features.
   b. Secondary Setup (if required): Rotation to access side holes or features.
g. Operation Sequence
h. Tool Selection:
   Tool Material: Carbide tools (optimized for aluminum).
i. Coolant: Flood the coolant to minimize heat and chip buildup.
j. Fixture and Clamping:
   First Setup: Standard vise or custom fixture for base.
   Second Setup: Rotated fixture with soft jaws for hole access.
k. Process Parameters:
   Spindle Speed: 6,000–12,000 RPM (depending on tool and operation).
   Feed Rate: 500–1,000 mm/min.
   Depth of Cut: Based on the tool and feature.
l. Post-Processing:
   G-code Generation: Output to CAM software.
m. Quality Control: An inspection plan (dimensions, tolerances) is generated.
n. Documentation: Route sheet and process plan PDF.

2.3.3 Process Planning for Parts

In part manufacturing, CAPP chooses the most economical route from raw material to finished product. It documents tool requirements, machining sequences, setup instructions, and quality control steps.

For example, Table 2.4 shows the route sheet for a custom flange which lists the raw stock as a forged disk, followed by turning, drilling, and facing operations with specific tools and machines noted.

TABLE 2.4

Route Sheet for a Custom Flange

Seq	Operation Description	Machine/Work Center	Tooling/Fixture	Estimated Time (min)	Remarks
10	Material preparation	Sawing machine	Material stopper	5	Cut blank from Ø210×20 mm rod
20	Facing (one side)	Lathe	3-Jaw chuck, facing tool	8	Flat surface for reference

(Continued)

TABLE 2.4 (*Continued*)

Route Sheet for a Custom Flange

Seq	Operation Description	Machine/Work Center	Tooling/Fixture	Estimated Time (min)	Remarks
30	OD and ID turning	Lathe	Boring bar, turning tool	12	OD: 200 mm, ID: 100 mm, ±0.05 mm
40	Drill center hole	Drill press	Ø10 mm center drill	5	Pilot for bolt pattern
50	Bolt hole drilling (6×Ø12 mm)	CNC machining center	Ø12 mm HSS drill, fixture plate	15	Holes on 150 mm PCD
60	Chamfer edges and holes	CNC or manual deburring	Chamfer tool	5	0.5×45° on holes and outer edge
70	Surface finish (top/bottom)	CNC mill or surface grinder	End mill or grinding wheel	10	Ra 1.6 μm finish on both faces
80	Final inspection	CMM and manual gauges	CMM probe, plug gauges	8	Check all critical dimensions
90	Cleaning and packaging	Workstation	Compressed air, poly wrap	5	Deburred, oil-free, labeled

OD, Outer Diameter; ID, Inner Diameter.

Example 2.3: CAPP for a Custom Flange

PART DESCRIPTION

A custom steel flange has the following features:

- Outer Diameter (OD): 200 mm
- Inner Diameter (ID): 100 mm
- Thickness: 15 mm
- 6 equally spaced bolt holes (Ø12 mm) on a 150 mm Pitch Circle Diameter (PCD)
- Material: AISI 1020 (low-carbon steel)
- Surface Finish: 1.6 μm on mating faces
- Tolerances: ±0.05 mm on critical dimensions

THE CAPP

The process plan is as shown in Table 2.5.

TOOLING INFORMATION

- Tools: Ø12 mm HSS drill, Ø10 mm center drill, facing tool, boring bar, chamfer tool, end mill for finish face
- Fixtures: Vise for drilling, chuck for turning
- Clamping: Precision soft jaws on lathe, custom fixture plate on CNC for hole drilling.

PROCESS PLANNING NOTES

- The CAD model of the flange is imported into the CAPP system.

TABLE 2.5
CAPP for a Custom Flange

Step	Operation	Machine/Tool	Parameters and Notes
1	Material selection	–	AISI 1020 steel blank Ø210×20 mm
2	Facing	Lathe+facing tool	Ensure flat reference surface
3	Turning (OD and ID)	Lathe+boring bar and external tool	OD to 200 mm, ID to 100 mm, tolerance ±0.05 mm
4	Drilling (center hole)	Drill press or CNC machining center	Drill a starter hole at the center
5	Bolt hole pattern drilling	CNC machining center+Ø12 mm drill	6 holes on 150 mm PCD, equally spaced, coordinates auto-calculated by CAD/CAPP
6	Chamfering	Manual or CNC – Chamfer tool	Apply chamfers on holes and edges (0.5×45°)
7	Surface finishing	Surface grinder or CNC face milling	Achieve 1.6 μm Ra on top and bottom faces
8	Deburring and inspection	Manual deburring+CMM	Clean and verify critical dimensions using CMM

- Feature recognition identifies holes, profiles, and faces for planning.
- The CAPP software selects operations based on feature geometry and material.
- Tool paths and cutting parameters (speeds and feeds) are suggested.
- Tolerances are linked to inspection and tool compensation routines.

2.3.4 Process Planning for Assemblies

Assemblies require planning for joining, sequencing, and resource allocation. Line balancing and task assignments are vital in such cases. For example, in an automated line assembling a motorbike engine, CAPP identifies the assembly order and allocates fastening tasks to robotic stations based on complexity and cycle time.

2.3.5 Decision-Making Methods in CAPP

Techniques like decision tables, AI algorithms, and fuzzy logic are used to assist process selection in CAPP. For example, a route sheet for a flange includes operations such as outer turning, internal boring, and facing, each tagged with a machine tool, speed, and inspection method. In CAPP, decision-making involves selecting the most appropriate manufacturing processes, sequences, tools, and machines based on part geometry, material, tolerances, and production volume. Modern CAPP systems incorporate a variety of decision-making methods to automate or support process planning.

2.3.5.1 Variant (Retrieval-Based) CAPP

This approach uses a database of standard process plans, which are retrieved and adapted for parts that belong to similar groups (as defined by group technology classifications). The decision for the retrieval and modification process is based on:

- Part family classification systems (e.g., Opitz or MICLASS codes)
- Predefined decision rules associated with each part family
- Manual edits by a human planner to refine the retrieved plan as needed

Example: A standard flange plan is retrieved and adapted to a custom flange by adjusting hole drilling and surface finish operations.

2.3.5.2 Generative CAPP

Using rule-based or algorithmic methods, generative systems create a new process plan from scratch based on part features, material, and other attributes. The following are the decision techniques used in Generative CAPP:

- Rule-Based Systems (if-then rules)
- Constraint Satisfaction (e.g., tool compatibility, tolerance limits)
- Decision Trees or Heuristics
- Knowledge-Based Expert Systems

Example: For a new component, the CAPP system selects operations, sequences, tools, and machines without human intervention by analyzing geometric and material data from the CAD model.

2.3.5.3 Semi-Generative (Hybrid) CAPP

Hybrid CAPP combines retrieval of standard plans (like variant CAPP) with generative modules to automate parts of the planning process. Its decision approach is based on:

- Human-in-the-loop decision-making
- Auto-generation of operations or tools while routing may be manually edited

Example: The system retrieves a turning plan for a cylindrical part and automatically inserts drilling operations based on recognized hole features.

2.3.5.4 Artificial Intelligence (AI)–Based Decision-Making

Modern CAPP systems apply AI techniques to optimize decisions, adapt to new data, and improve over time. The methods used are as follows:

- Machine Learning: Predicts optimal tool paths or machines based on historical data
- Fuzzy Logic: Handles uncertainty in machining conditions (e.g., surface finish vs. tool wear)
- Neural Networks: Maps complex relations between part features and process sequences
- Case-Based Reasoning (CBR): Learns from past planning cases.

Example: A neural network trained on aerospace components predicts an optimal sequence of operations and machines for a new titanium part.

2.3.5.5 Multi-Criteria Decision-Making (MCDM)

MCDM is used when multiple conflicting objectives (cost, quality, time) need to be balanced. MCDM methods include:

- Analytic Hierarchy Process (AHP): Decomposes decisions into a hierarchy of goals and alternatives
- Technique for Order of Preference by Similarity to Ideal Solution (TOPSIS): Ranks alternatives by measuring their relative closeness to the ideal and farthest solutions based on multiple criteria
- Weighted Sum Models: Scores process options against prioritized criteria.

Example: CAPP selects between two alternative milling sequences by scoring them based on cost, tool life, and machining time.

2.3.5.6 Optimization-Based Planning

It formulates the process plan as a mathematical optimization problem (e.g., minimizing cost or cycle time). The techniques used are:

- Linear Programming (LP)
- Genetic Algorithms (GA)
- Simulated Annealing
- Dynamic Programming.

TABLE 2.6

Comparison Table for Decision-Making Methods in CAPP

Method	Automation Level	Flexibility	Intelligence	Example Use Case
Variant	Low	Medium	Rule-based	Standard part families
Generative	High	High	Rule/logic	New parts, high variation
AI-based	High	High	Learning	Complex, data-rich environments
MCDM	Medium	Medium	Evaluation	Conflicting criteria decisions
Optimization-based	High	Medium	Mathematical	Cost/time optimization

Example: GA finds the best combination of machines and sequences for batch production to minimize material movement and setup time.

Table 2.6 shows the comparison table for decision-making methods in CAPP.

2.3.6 Integration of CAPP with Other Systems

CAPP can be integrated with other systems such as CAD, CAM, and ERP/ Manufacturing Execution System (MES).

- **CAD → CAPP**: Automatic geometric and feature data extraction can be done.
- **CAPP → CAM**: Direct transfer of process plans is done to generate NC toolpaths.
- **CAPP → ERP/MES**: Routing, scheduling, and BOM data transfer are done for execution and resource planning.

2.4 Computer-Aided Manufacturing (CAM)

CAM is the use of computer software and hardware to control machine tools and automate the manufacturing of products. CAM leverages digital data, often originating from CAD and CAPP systems, to precisely dictate machining strategies, tooling paths, and cutting operations. This results in high-speed, high-precision manufacturing with reduced waste and improved product consistency.

2.4.1 Functions and Capabilities of CAM

a. Toolpath Generation: CAM systems convert geometric models into a series of tool movements that a CNC machine can follow.
b. Post-Processing: CAM software translates generic toolpaths into specific machine language (e.g., G-codes) based on the controller used.
c. Simulation and Verification: These functions simulate machining operations to detect collisions, validate the tool path, and estimate machining time.
d. Optimization: CAM software can recommend or automatically adjust feed rates, tool sequences, and paths for optimal efficiency.

2.4.2 CAM in Various Manufacturing Processes

CAM plays a crucial role across various types of manufacturing processes, each leveraging its capabilities in distinct ways. In subtractive manufacturing, which includes operations like milling, turning, drilling, and grinding, CAM generates precise tool paths that remove material from a solid block to form the desired geometry. With the advancement of additive manufacturing, CAM systems have evolved to support processes such as 3D printing by generating slicing paths and managing layer-by-layer deposition strategies. Furthermore, hybrid manufacturing, which integrates additive and subtractive processes in a single machine, also benefits from CAM software that coordinates complex tool changes, motion paths, and process sequencing within the same workflow. These capabilities make CAM an essential component in modern and emerging manufacturing environments.

2.4.3 Integration with CNC

CAM systems are essential for CNC machining. They output programs tailored to the exact kinematics and limitations of CNC machines. High-speed machining (HSM) requires precise toolpaths and adaptive control to maintain part integrity at high feed rates. CAM software helps reduce machine wear, improve surface finish, and decrease cycle time.

2.4.4 G-Codes and M-Codes in CNC Programming

Geometric codes (G-codes) and Miscellaneous codes (M-codes) are fundamental to CNC programming. They instruct CNC machines on how to move, operate, and interact with the workpiece, tooling, and auxiliary functions.

2.4.4.1 G-Codes: Motion and Positioning Commands

G-codes direct the machine's tool path, defining how the cutting tool moves in relation to the workpiece. Table 2.7 shows the common G-codes with examples.

2.4.4.2 M-Codes: Machine Function Commands

M-codes control non-motion functions, such as starting/stopping the spindle, coolant, and program flow. Table 2.8 shows the common M-codes with examples.

TABLE 2.7

Common G-Codes with Examples

Code	Function	Example Command	Description
G00	Rapid positioning	G00 X100 Y50	Moves the tool rapidly to $X=100$, $Y=50$ without cutting
G01	Linear interpolation (cutting move)	G01 X50 Y25 F150	Cuts a straight path to $X=50$, $Y=25$ at feed rate 150 mm/min
G02	Circular interpolation clockwise	G02 X50 Y50 I25 J0	Moves in a clockwise arc ending at $X=50$, $Y=50$, with arc center offset $I=25$, $J=0$
G03	Circular interpolation counterclockwise	G03 X10 Y10 I0 J-10	Moves in a counterclockwise arc
G17	XY plane selection	G17	Sets XY plane for circular motion commands (default for milling)
G20	Programming in inches	G20	Switches all dimensioning to inches
G21	Programming in millimeters	G21	Switches all dimensioning to millimeters
G28	Return to the machine home position	G28 U0 W0	Moves axes U and W to machine zero position
G90	Absolute positioning	G90	Positions are given with respect to the origin
G91	Incremental positioning	G91	Positions are given relative to the current position

TABLE 2.8

Common M-Codes with Examples

Code	Function	Example Command	Description
M00	Program stop	M00	Stops the machine; resumes manually
M01	Optional stop	M01	Stops if optional stop is active on the control panel
M02	End of program	M02	Ends the CNC program

(*Continued*)

TABLE 2.8 (*Continued*)

Common M-Codes with Examples

Code	Function	Example Command	Description
M03	Spindle on (clockwise)	M03 S1200	Starts the spindle clockwise at 1,200 RPM
M04	Spindle on (counterclockwise)	M04 S800	The spindle starts counterclockwise at 800 RPM
M05	Spindle stop	M05	Stops the spindle
M06	Tool change	M06 T2	Changes to tool number 2
M08	Coolant on	M08	Turns the coolant on
M09	Coolant off	M09	Turns the coolant off
M30	End of program and reset	M30	Ends program and resets CNC controller to start

2.4.5 Digitizing Manufacturing

CAM systems use CAD/CAPP outputs to generate toolpaths and control machine tools via CNC. CAM improves consistency, reduces waste, and enables high-precision manufacturing. For example, a part with multiple slots and pockets is machined using G-codes generated from CAM software, which controls a CNC mill to follow the defined tool paths.

2.4.6 CNC, DNC, and FMS

CNC, DNC and FMS are integral parts of CAM. CNC executes machine instructions from a digital program, while DNC links multiple CNCs to a central computer. FMS integrates machines with automated material handling for mid-volume, mid-variety production. For example, in a furniture factory, a DNC system synchronizes multiple CNC routers cutting various wood components from a centralized control unit.

2.4.7 Computer-Aided Numerical Control Programming (NCP)

NCP forms the bridge between process planning and machine execution. The programming process defines how a machine will carry out specific operations using manual input, computer assistance, or interactive graphics systems. It involves various types of machining centers (vertical and horizontal) that can perform various operations like milling, drilling, and tapping under programmed control. STEP-NC is an advancement over traditional G-code that enables intelligent machining by embedding design and manufacturing data, allowing CNC machines to interpret feature-based instructions directly.

Example: A STEP-NC enabled system recognizes a hole feature and automatically selects the right tool and parameters for drilling, adjusting in real-time if necessary.

2.4.7.1 Numerical Control Programming

NC programs use G-codes for movement (e.g., G01 for linear motion) and M-codes for miscellaneous functions (e.g., M03 for spindle on). The types of programming are as shown below:

- Manual Programming: It involves manually writing code for each operation, which is useful for simple parts or one-off jobs. For instance, an operator writes an NC program for a turned shaft with facing, rough turning, and threading operations.
- Computer-Assisted Part Programming: It uses software to automate code generation from part geometry and machining parameters. For example, CAM software reads a 3D model and outputs a toolpath with speeds and feeds tailored to the material.
- Interactive Graphics NC Programming: It enhances usability by letting programmers visually simulate and edit machining operations. For example, a user interacts with a 3D model of a mold cavity, selects surfaces to machine, and the software generates optimized G-code.

2.4.8 CAM and Additive Manufacturing (AM)

The rise of AM, such as 3D printing, has expanded the scope of CAM systems. Advanced CAM software now includes slicing algorithms, support structure generation, and process parameter optimization for AM processes.

Example: A lattice-structured bracket is designed in CAD and processed through CAM software that determines the layer-by-layer toolpath for a selective laser melting (SLM) machine.

2.4.9 Simulation and Verification in CAM

Modern CAM systems often include simulation modules to verify machining operations before they are executed. This helps avoid collisions, optimize tool paths, and reduce scrap.

Example: A multi-axis milling operation is simulated in a digital twin environment to validate the tool motion and detect potential collisions with fixtures or clamps. Once verified, the program is sent directly to the machine for execution.

2.5 CAD–CAPP–CAM Integration with Intelligent Manufacturing

The synergy between CAD, CAPP, and CAM minimizes errors and enhances productivity. Modern systems use feature-based modeling, APIs, and database sharing to automate the flow from design to shop floor. For instance, an intelligent CAPP platform extends SolidWorks using APIs to auto-identify machining features and generate toolpaths for a hydraulic press brake component. Figure 2.4 shows the entire production process across CAD, CAPP, and CAM.

Moreover, when coupled with intelligent manufacturing, each of CAD, CAPP, and CAM has a unique role. Intelligent CAD will generate intelligent design systems with generative design, design-rule checking, real-time simulation, and optimization integrated with sensor feedback (via intelligent CAE). With respect to CAPP, dynamic process planning will be possible which can adapt to real-time machine or resource statuses. Moreover, smart toolpath generation based on real-time feedback from CNC systems will be the result of intelligent CAM systems.

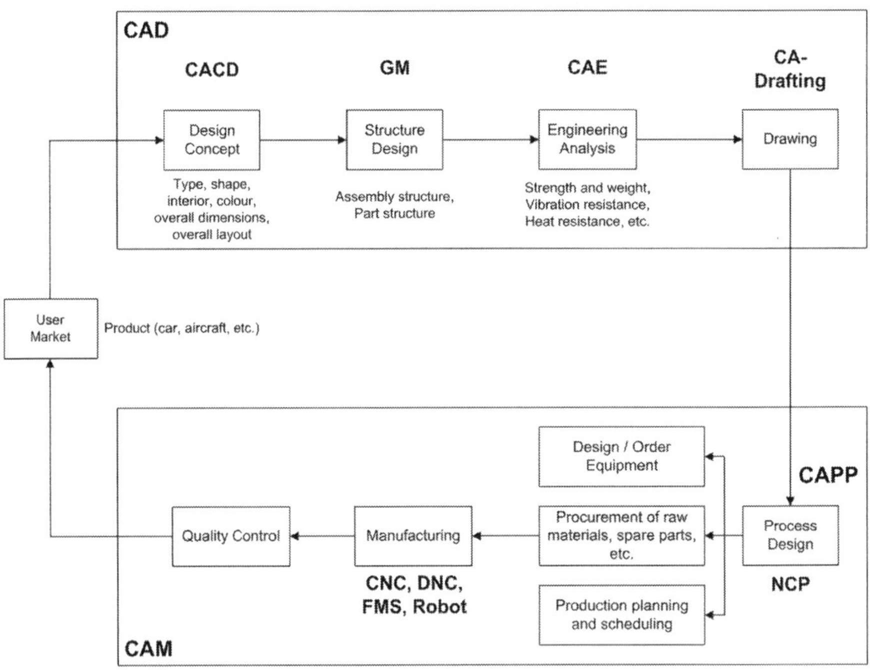

FIGURE 2.4
Entire production process.

2.5.1 Future Trends in CAD

The future of CAD is being shaped by several emerging technologies that enhance creativity, collaboration, and efficiency. AI-assisted design streamlines the development process by using generative design algorithms to propose optimized design alternatives based on functional requirements, significantly reducing design cycles. For instance, Autodesk Fusion 360 can automatically generate geometry optimized for strength and material efficiency.

Cloud-based and collaborative CAD platforms are also gaining traction, allowing teams to co-edit models in real-time, maintain version control, and access projects from any device. Examples include Onshape and Fusion 360.

Another significant advancement is the integration of CAD with simulation tools (CAE), enabling real-time stress, thermal, and dynamic analyses during the design phase. This shift toward "simulation-driven design" is rapidly becoming standard practice.

Furthermore, extended reality (XR) technologies, such as augmented and virtual reality, are integrated to enhance spatial understanding and provide immersive design reviews. Lastly, parametric and rule-based design automation is expanding through tools like Grasshopper for Rhino and Inventor iLogic, enabling rapid design variations and automating repetitive tasks based on predefined rules and parameters.

2.5.2 Future Trends in CAPP

The future of CAPP is moving toward greater automation, intelligence, and integration. Powered by AI, fully generative and autonomous CAPP systems will automatically generate optimized process plans based on CAD data, production constraints, and historical performance. These systems will be tightly integrated with enterprise solutions such as ERP and MES, enabling continuous improvement through closed-loop feedback.

Feature recognition with semantic understanding will enhance the ability of CAPP systems to intelligently interpret CAD models, allowing for context-aware mapping of features to suitable manufacturing operations. Another major trend is real-time adaptive planning, where CAPP systems will dynamically modify process plans based on live shop floor conditions such as machine availability, tool wear, or material changes.

Additionally, cloud-based process planning will facilitate centralized process data management across global facilities, enabling standardization and rapid dissemination of best practices. Finally, the rise of knowledge-based and learning systems will allow CAPP platforms to evolve by learning from historical plans, operator feedback, and production results, ultimately leading to more accurate and efficient planning.

2.5.3 Future Trends in CAM

AI-driven toolpath optimization will use machine learning to predict the most efficient toolpaths, cutting parameters, and machining strategies tailored to specific part geometries and materials. Digital twin integration will allow CAM systems to simulate and optimize machining processes within virtual replicas of machines and production lines, reducing errors and setup times.

Closed-loop manufacturing is another key trend, where real-time data from CNC machines, such as spindle load and vibration, will be fed back into CAM systems to adjust machining parameters dynamically. Furthermore, CAM for additive and hybrid manufacturing is gaining prominence, enabling support for technologies like directed energy deposition and multi-axis 3D printing alongside traditional subtractive processes.

The shift to cloud-based CAM will facilitate remote toolpath generation and machining operations, providing flexibility and scalability, especially for small and medium-sized enterprises. Lastly, with increased connectivity, cybersecurity and data integrity will become essential, ensuring that NC codes and toolpath data are securely transmitted and protected against tampering or loss.

2.6 Highlights

The highlights of the chapter are as follows:

- Comprehensive coverage of CAD, including conceptual, geometric, parametric, and feature-based modeling for engineering design.
- Detailed discussion on CAPP workflows, part and assembly planning, decision-making methods (rule-based, AI, MCDM, and optimization), and CAD, CAM, and ERP/MES integration.
- Thorough explanation of CAM systems, including toolpath generation, CNC programming (G-codes and M-codes), simulation, and support for subtractive, additive, and hybrid manufacturing.
- Case examples and route sheets provided for prismatic parts and flanges, demonstrating real-world application of CAPP and CAM planning techniques.
- Integration framework showing seamless data flow and process linkage between CAD, CAPP, and CAM systems for intelligent manufacturing.
- Future outlook highlights via AI-driven generative design, real-time adaptive CAPP, and digital twin-enabled CAM for Industry 4.0 and beyond.

Exercises

• CAD

P2.1 What is Computer-Aided Design (CAD)? Describe its main objectives in product development.

P2.2 Differentiate between wireframe, surface, and solid modeling. Give one application for each.

P2.3 Explain the advantages of using parametric modeling in modern CAD systems.

P2.4 What is geometric constraint in CAD? Provide examples.

P2.5 Briefly explain the role of layers in a 2D CAD environment.

P2.6 Draw and dimension a 2D sketch of a mechanical component using any CAD software.

P2.7 List and explain at least five types of geometric entities used in CAD modeling.

P2.8 Discuss how CAD systems support collaborative design in engineering teams.

P2.9 Explain the difference between part modeling and assembly modeling. Why is this distinction important?

P2.10 What is feature-based modeling? How does it help in design modification?

P2.11 Create a 3D model of a bracket with at least one hole, boss, and fillet using your preferred CAD software. Submit screenshots.

P2.12 Use a CAD tool to simulate an exploded view of a simple assembly. Annotate key components.

P2.13 Design a flange in 2D, then extrude it to 3D. Include constraints and dimensions in your submission.

P2.14 Export a part model to STL and describe the process. Why is STL important in CAD-CAM integration?

P2.15 Analyze the difference in file formats: .DWG, .DXF, .IGES, and .STEP. What are their use cases?

• CAM

P2.16 What is Computer-Aided Manufacturing (CAM)? How does it relate to CAD and CAPP?

P2.17 Define G-code and M-code. What are their roles in CNC programming?

P2.18 Explain the importance of post-processing in CAM.

P2.19 What is toolpath generation? What factors influence the choice of a toolpath?

P2.20 Differentiate between 2½-axis, 3-axis, and 5-axis machining. Provide examples.

P2.21 List and explain at least five benefits of using CAM in modern manufacturing.

P2.22 What is the role of simulation in CAM software? How does it help in error detection?

P2.23 Describe the steps involved in generating NC code for a simple prismatic part.

P2.24 Explain the significance of cutting parameters such as feed rate, spindle speed, and depth of cut.

P2.25 Discuss the concept of tool libraries in CAM software. Why are they important?

P2.26 Using any CAM software, generate the toolpath for milling a rectangular pocket. Submit screenshots of the toolpath and final simulation.

P2.27 Create a basic CNC turning operation (e.g., facing and turning) for a cylindrical part.

P2.28 Simulate a machining process using CAM and identify at least two potential issues (e.g., tool collision, excess stock).

P2.29 Perform a post-processing operation and explain how the output G-code differs for two types of CNC machines (e.g., Fanuc vs. Siemens).

P2.30 Design a simple fixture for a given part and describe how it would be used during machining.

- **CAPP**

P2.31 What is Computer-Aided Process Planning (CAPP)? How does it bridge CAD and CAM?

P2.32 Differentiate between Variant CAPP and Generative CAPP. Give an example of each.

P2.33 List the key components of a process plan.

P2.34 Explain the role of feature recognition in CAPP.

P2.35 What are the benefits of using CAPP in a manufacturing environment?

P2.36 Describe the typical steps involved in generating a process plan from a CAD model.

P2.37 How does CAPP help in reducing lead time and cost in manufacturing?

P2.38 Explain the importance of selecting the correct machining operations and sequencing them effectively.

P2.39 Discuss how tooling and fixture selection are influenced by process planning.

P2.40 What data is required from a CAD model to effectively perform process planning?

3

Intelligent Control in Manufacturing Systems

3.1 Introduction

As industrial automation continues to evolve, there's a major push toward digital transformation, driven by technologies like cyber-physical systems, the IoT, and AI. These technologies are changing how factories operate, with emphasis on real-time responsiveness, data-driven decision-making, and interconnectivity among machines, systems, and humans. As manufacturing systems become complex, they demand control strategies that are accurate and reliable, adaptive, self-learning, and capable of making decisions in dynamic environments.

Though effective in static and well-modeled settings, traditional control systems often are limited in real-time applications involving nonlinearity, high-dimensional data, and uncertainty. These limitations become more pronounced as manufacturers strive for mass customization, reduced downtime, and operational resilience under unexpected disruptions such as machine failure or supply chain variability. As a result, there is an increasing need for intelligent and autonomous systems.

Intelligent control systems have emerged to address these challenges by integrating AI and ML techniques with traditional control practices. They mimic human decision-making and adaptivity through data-driven models, self-organizing logic, and heuristic algorithms, enabling real-time optimization, fault tolerance, and autonomous behavior.

This chapter presents the development from traditional to intelligent control systems, introducing the basic terminologies, operational principles, and various intelligent techniques that form the basis of modern control strategies in IM. It provides a foundation for understanding how advanced control methodologies can be designed, implemented, and scaled to meet the demands of next-generation industrial environments.

 DOI: 10.1201/9781003485186-4

3.2 Traditional and Distributed Control Approaches

3.2.1 Traditional Control Systems

Traditional control systems are grounded in control theory, which models the behavior of dynamic systems using differential equations and state-space representations. These systems are generally implemented using techniques such as:

- Proportional-Integral-Derivative (PID) Control: One of the most commonly used feedback control methods, PID control regulates processes by continuously calculating an error value and applying a corrective action (a block diagram of a PID controller is shown in Figure 3.1).

  Figure 3.1 shows a PID controller, in which an error value $e(t)$ is continuously calculated as the difference between a desired setpoint $r(t)$ and the measured process variable $y(t)$. This value is then corrected based on proportional (P), integral (I), and derivative terms (D). The controller minimizes the error over time by adjusting the control variable $u(t)$.
- Model Predictive Control (MPC): This technique uses a model of the system to predict future behavior and optimize control moves accordingly.

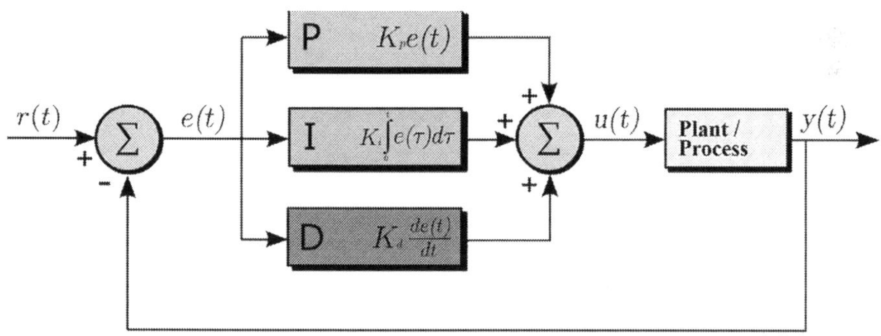

FIGURE 3.1
Block diagram of PID controller. (By Arturo Urquizo – File:PID.svg, CC BY-SA 3.0, https://commons.wikimedia.org/w/index.php?curid=17633925.)

- State-Space Control and Pole Placement: These are useful for multiple-input multiple-output (MIMO) systems and involve designing feedback that drives system states to desired values.

Key characteristics include:

- Determinism: System behavior is entirely defined by known equations.
- Centralization: Control computations and decision-making occur at a central location.
- Limited Adaptability: Changes in system dynamics or disturbances require manual tuning or redesign of the controller.

Despite their strengths in predictability and mathematical rigor, traditional systems struggle to cope with unforeseen conditions, sensor noise, nonlinear behavior, and uncertainties frequently arising in modern manufacturing setups.

3.2.2 Distributed Control Systems (DCS)

Distributed control systems have been developed to address the scalability and reliability limitations of centralized systems. DCS architectures consist of multiple controllers distributed across the plant floor, each managing specific subsystems or machines. These controllers communicate via industrial networks such as Ethernet/IP, Profibus, or Modbus.

Features of DCS include:

- Local autonomy where each node can make decisions based on localized information.
- Inter-node communication where information sharing enables coordination across subsystems.
- Improved Fault Tolerance: Failures in one part of the system do not necessarily compromise the entire system.
- Hierarchical Control: These systems are often built-in layers at different levels, such as field, control, and supervisory levels, so that each layer can focus on a specific part of the process. This makes it easier to modify and control the performance at different levels.

While DCS significantly improves robustness, scalability, and responsiveness, it relies on rule-based logic or model-based algorithms that do not learn or adapt beyond their initial programming. As industrial processes become increasingly complex, these systems must be enhanced with intelligence to handle variability, optimize operations autonomously, and learn from experience.

3.3 Intelligent Control Architecture

Intelligent control architecture is designed to embed cognitive capabilities into automated systems. Unlike traditional control, which is reactive and dependent on predefined models, intelligent control can reason, learn, and adapt in real time. It draws inspiration from human cognitive functions and biological systems, applying them to engineering problems.

3.3.1 Core Components

1. Perception Layer:
   - Serves as the system's sensory input, gathering data from physical processes via sensors.
   - Performs preprocessing such as denoising, normalization, and feature extraction to convert raw signals into useful information.
   - Incorporates machine vision, audio sensors, vibration analysis, and other modalities in complex manufacturing settings.
2. Knowledge Layer:
   - Acts as a memory and inference base. It stores historical data, learned patterns, heuristics, and expert rules.
   - Maintains a representation of the environment and system state using models, logic, and structured knowledge bases.
   - Can be updated through supervised learning, unsupervised clustering, or human input.
3. Decision Layer:
   - Functions as the cognitive core. It uses machine learning algorithms (e.g., neural networks, decision trees, and reinforcement learning) to infer the best control actions based on current data.
   - Capable of planning, goal setting, and adapting to new situations with minimal human intervention.
   - May include predictive modules that anticipate future conditions or disruptions.

4. Execution Layer:
    - Interfaces with actuators and process controllers to carry out decisions made by the system.
    - Employs real-time feedback control to ensure that the desired outcomes are achieved.
    - Often includes safety and fallback mechanisms for error detection and recovery.

3.3.2 Design Principles

- Modularity: Allows each layer to be developed and updated independently while maintaining a coherent overall structure.
- Adaptivity: Enables the system to respond to environmental changes and learn over time without explicit reprogramming.
- Robustness: Maintains performance despite disturbances, noise, or uncertainties.
- Interoperability: Integrates easily with existing industrial communication protocols and legacy systems.
- Transparency and Explainability: Important in critical systems where operators must understand decision logic.

3.3.3 Architectural Variants

Several architectural paradigms are being explored and applied in intelligent control systems:

- Hybrid Intelligent Systems: These combine symbolic AI (e.g., rule-based systems) with sub-symbolic approaches (e.g., neural networks) to leverage the strengths of both.
- Multi-Agent Systems (MAS): In MAS, independent agents cooperate and communicate to achieve collective goals. Each agent has autonomy and may represent a machine, sensor cluster, or function.
- Edge and Cloud Integration: Edge computing enables low-latency decision-making at the machine level, while cloud computing offers high-capacity data processing and long-term learning models.
- Digital Twins: These are virtual representations of physical systems used for simulation, optimization, and predictive analytics. They can be tightly integrated with the intelligent control architecture.

In summary, intelligent control architecture transforms how automated systems perceive, reason, and act. By merging AI with control theory, these systems enable factories to transition from reactive control to proactive, predictive, and autonomous operation. This paradigm boosts productivity and efficiency and fosters innovation by allowing flexible, resilient, and human-aware industrial environments.

3.4 Types of Intelligent Control Techniques

In IM systems, various intelligent control techniques utilize advanced algorithms and AI principles to adapt, optimize, and make autonomous decisions. These techniques move beyond the limitations of traditional control methods, enabling systems to learn from experience, handle uncertainties, and make decisions in real time. Below is a detailed explanation of the key intelligent control techniques:

3.4.1 Rule-Based Systems

3.4.1.1 Basic Definition and Terminology

Rule-based systems, also known as expert systems, are designed to emulate human decision-making using sets of "if-then" rules. These rules are based on expert knowledge or heuristics, which define how specific conditions lead to certain actions. The system evaluates inputs and determines outputs based on predefined logical structures.

Key terminology

- Rules: The fundamental building blocks of rule-based systems. Each rule consists of two parts: the condition (the "if" part) and the action (the "then" part). For example, "If temperature > 80°C, then activate cooling system."
- Knowledge Base: This collects all the rules, facts, and data the system uses to make decisions. It represents expert knowledge about the domain.
- Inference Engine: The component that interprets the rules and applies them to the facts or input data to draw conclusions. It runs the reasoning process to apply the "if-then" rules.
- Forward Chaining: It is a process where the inference engine starts with known facts and applies rules to infer new facts. It is data-driven.
- Backward Chaining: It is a process where the system starts with a goal or hypothesis and works backward to see if the rules can support the hypothesis. It is goal-driven.

Rule-based systems are widely used in areas where rules are clear and can be easily encoded. However, they cannot learn or adapt to new situations unless new rules are explicitly added.

3.4.1.2 Applications in Intelligent Manufacturing

In IM, rule-based systems can automate tasks that require expert knowledge and decision-making. Common applications include:

- Quality Control: Rules can be created to detect defects in materials or products, such as "If the diameter of a part is out of tolerance, then reject the part."
- Fault Diagnosis: Rules can be used to identify failures in machinery or processes. For instance, "If the machine is vibrating above a certain threshold, then check for mechanical failure."
- Process Regulation: Rules can be used to control parameters like temperature, pressure, or speed in manufacturing processes.

3.4.1.3 Examples

A common example is an expert system used in predictive maintenance. The system might have rules such as:

- "If machine X has been running for more than 1,000 hours, then alert for scheduled maintenance."
- "If sensor Y reports abnormal readings, then initiate diagnostic procedures."

3.4.2 Decision Trees

3.4.2.1 Basic Definition and Terminology

A decision tree is a supervised learning model that maps observations about an item to conclusions. It is used for classification and regression tasks. The structure is tree-like, with nodes representing decisions, branches showing possible outcomes, and leaf nodes representing the final classification or value.

Key terminology

- Root Node: The top node of the tree that represents the entire dataset and is split into subsets based on the feature with the most significant value.

- Internal Nodes: These represent decision points where the dataset is split according to certain attributes or features.
- Branches: The connections between nodes that represent the outcome of a decision.
- Leaf Nodes: The final output nodes, where the classification or regression result is obtained.
- Splitting: The process of dividing a dataset into subsets based on the value of an attribute.
- Pruning: The process of removing branches from the tree that do not provide useful information, thus reducing overfitting.

Decision trees work by recursively partitioning the data into subsets based on the most significant attributes to achieve maximum classification purity in each subset.

3.4.2.2 **Applications in Intelligent Manufacturing**

In manufacturing, decision trees are used for tasks that involve classification or decision-making based on various input features. Examples include:

- Defect Detection: Decision trees can classify parts as defective or non-defective based on measurements such as size, weight, and surface condition.
- Supply Chain Optimization: Decision trees help make decisions about inventory management, such as "If demand increases by 20%, then restock inventory."
- Process Monitoring: Monitoring the health of machinery by analyzing sensor data and classifying the machine's status.

3.4.2.3 **Examples**

An example could be a decision tree used for quality assurance in a manufacturing process. The tree might classify products based on several characteristics, such as:

- Is the diameter within specification?
- Is the surface free of cracks?
- Is the weight within range?

Each decision point splits the data, and the leaves determine if the product is "Pass" or "Fail."

3.4.3 Evolutionary Computation Algorithms

3.4.3.1 Basic Definition and Terminology

Evolutionary computation algorithms are optimization techniques inspired by the process of natural evolution. They are used to find approximate solutions to complex problems, especially when the solution space is large and poorly understood. The most common evolutionary algorithms include GA, Genetic Programming (GP), and Particle Swarm Optimization (PSO).

Key terminology

- Population: A set of potential solutions (individuals) evolved over generations.
- Fitness Function: A function that evaluates how good a solution is. The higher the fitness, the better the solution.
- Selection: The process of choosing individuals based on their fitness to produce offspring for the next generation.
- Crossover: A genetic operator that combines parts of two solutions (parents) to create a new solution (offspring).
- Mutation: A process that introduces random variations into offspring to explore new areas of the solution space.
- Generations: A cycle of evolution in which a population undergoes selection, crossover, and mutation to evolve better solutions.
- Elitism: The practice of carrying the best solutions from one generation to the next without modification.

Evolutionary algorithms are often used to solve optimization problems where traditional methods may struggle, especially in nonlinear or multi-objective contexts. To have a general idea, the steps of GA are shown in Figure 3.2.

3.4.3.2 Applications in Intelligent Manufacturing

In manufacturing, evolutionary algorithms are used to optimize complex processes and systems. Applications include:

- Scheduling and Resource Allocation: Evolutionary algorithms can optimize production schedules, minimize downtime, or balance workloads across multiple machines.
- Process Design: They can be used to optimize the design parameters of processes, such as the speed of a robotic arm or the trajectory for part handling.
- Supply Chain Optimization: These algorithms help in routing, inventory management, and logistics optimization.

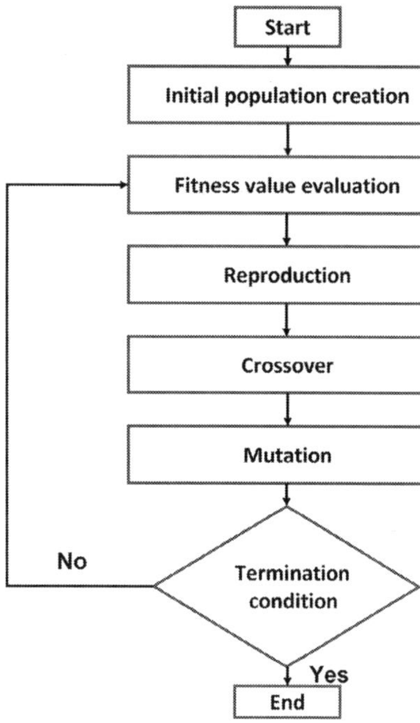

FIGURE 3.2
Block diagram of GA. (Adapted from El Houby, E. M., & Yassin, N. I. (2020). Wavelet-Hadamard based blind image watermarking using genetic algorithm and decision tree. *Multimedia Tools and Applications*, *79*(37), 28453–28474.)

3.4.3.3 Examples

For instance, a GA could be applied to optimize the path of a robot in a warehouse to minimize travel time. The algorithm would evolve solutions over several generations, with fitness being defined as the total time taken for the robot to complete its task.

3.4.4 Artificial Neural Networks (ANNs)

3.4.4.1 Basic Definition and Terminology

Artificial Neural Networks (ANNs) are computational models inspired by the structure and functioning of the human brain. They consist of layers of neurons (also called nodes) that process and transmit information. ANNs are used for various tasks, such as pattern recognition, classification, and regression.

Key terminology

- Neurons: Basic units of an ANN that process input data and produce an output. Each neuron has a set of weights that modify the input.

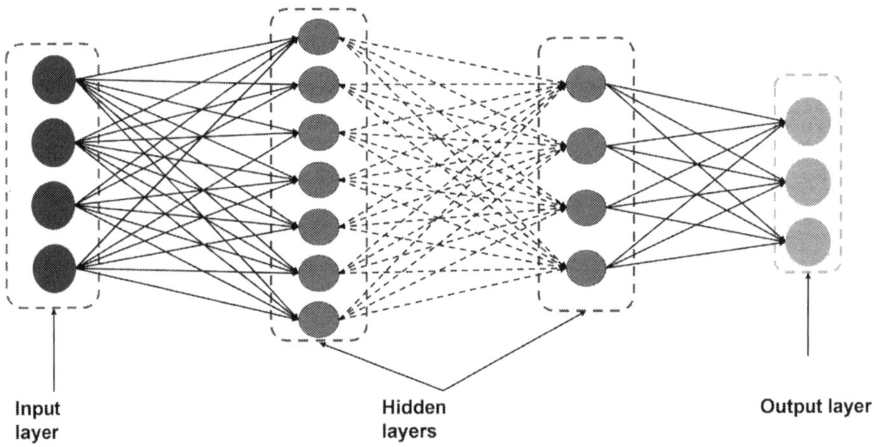

FIGURE 3.3
Block diagram of ANN architecture. (Adapted from Sadek, S., Al-Hamadi, A., Michaelis, B., & Sayed, U. (2011). Face detection and localization in color images: An efficient neural approach. *Journal of Software Engineering and Applications*, 4(12), 682.)

- Layers: ANNs have different layers (as shown in Figure 3.3):
  - Input Layer: The first layer receives the raw data.
  - Hidden Layers: Intermediate layers that process the data and perform calculations.
  - Output Layer: The final layer that produces the results.
- Activation Function: A function applied to the output of each neuron to introduce nonlinearity. Standard functions include sigmoid, ReLU, and tanh.
- Weights and Biases: Parameters that determine the strength of connections between neurons and help in learning from the data.
- Backpropagation: A learning algorithm used to adjust the weights and biases in the network by propagating errors backward through the network.

ANNs can adapt to new data, making them suitable for classification, regression, and predictive modeling tasks.

3.4.4.2 Applications in Intelligent Manufacturing

In manufacturing, ANNs are widely used for:

- Predictive Maintenance: Predicting when equipment will fail based on sensor data, operational history, and patterns.

- Fault Detection: Identifying anomalies or deviations from normal behavior that could indicate a problem.
- Process Optimization: Adjusting process parameters in real time to optimize performance.

3.4.4.3 Examples

An example could be using an ANN to predict the remaining useful life of a machine. The network might take inputs like vibration data, temperature, and operational hours, and output the likelihood of failure within the next 'X' days.

3.4.5 Fuzzy Logic

3.4.5.1 Basic Definition and Terminology

Fuzzy logic is a mathematical framework that deals with uncertainty and imprecision. Unlike traditional binary logic (true/false), fuzzy logic allows for degrees of truth, meaning that values can range between 0 and 1. This is particularly useful when dealing with vague or imprecise information.

Key terminology

- Fuzzy Sets: These are sets where elements have degrees of membership, ranging from 0 (not belonging) to 1 (completely belonging). For example, "temperature is hot" could have a membership of 0.7, meaning it is somewhat hot.
- Membership Functions: These functions map inputs to values between 0 and 1, defining the degree to which an element belongs to a fuzzy set.
- Rule Base: A set of "if-then" rules used to perform reasoning in fuzzy logic systems. These rules can deal with linguistic variables like "high," "low," or "medium."
- Fuzzy Inference System (FIS): The system that performs fuzzy reasoning based on the fuzzy rules. It involves three steps: fuzzification (converting crisp inputs into fuzzy values), rule evaluation (applying fuzzy rules), and defuzzification (converting fuzzy outputs back into crisp values).

3.4.5.2 Applications in Intelligent Manufacturing

Fuzzy logic is used in manufacturing where precise values are difficult to obtain. Applications include:

- Temperature Control: Adjusting heating elements based on temperature, where precise measurements are difficult to define (e.g., "temperature is high" could mean slightly high, moderately high).
- Speed Control: Fuzzy controllers can be used in conveyor systems to adjust speed based on load or other varying conditions.
- System Diagnosis: Handling complex scenarios with uncertain data to diagnose faults or predict failures.

3.4.5.3 Examples

An example of fuzzy logic in manufacturing might be a controller for a CNC machine. The fuzzy logic system could adjust the cutting speed based on variables such as tool wear and vibration, where exact numerical thresholds may not be easily defined.

3.5 Highlights

- Intelligent Control Systems: The chapter discusses the transition from traditional control systems to intelligent control systems, driven by the need for more adaptive, flexible, and autonomous manufacturing systems in Industry 4.0. Traditional control techniques, like PID control and MPC, are often insufficient in handling dynamic, complex, and uncertain environments. Intelligent control systems, by contrast, leverage AI, machine learning, and data-driven models to make decisions in real time, enabling better performance, efficiency, and fault tolerance.
- Key Technologies in Intelligent Control: Several AI-based techniques, such as fuzzy logic, decision trees, neural networks, and evolutionary algorithms, are fundamental to intelligent control systems. These methods enable machines to adapt, optimize processes, and predict outcomes, significantly improving productivity and reducing downtime.
- DCS vs Intelligent Control: The chapter compares traditional and DCS with intelligent control systems. While DCS offers localized control with better fault tolerance and scalability, it lacks the adaptability and decision-making capabilities of intelligent systems. Intelligent control systems, which use data and learning algorithms, provide the flexibility needed to handle nonlinearity, high-dimensional data, and complex manufacturing environments.

- Architecture of Intelligent Control Systems: The intelligent control architecture is made up of four main layers: perception, knowledge, decision, and execution. Each of these layers is responsible for gathering information, processing and storing it, making decisions, and taking actions accordingly. This layered structure enhances the system's ability to respond to changes and make predictive adjustments autonomously.
- Techniques for Intelligent Control: Several intelligent control techniques are discussed:
  - Rule-Based Systems: These systems use expert rules to make decisions based on "if-then" logic.
  - Decision Trees: These are used to model decisions and classify outputs based on input data.
  - Evolutionary Algorithms: Techniques like genetic algorithms are used for optimization, such as determining the best parameters for system performance.
  - ANNs: These are used for pattern recognition, fault detection, and predictive maintenance by learning from large datasets.
  - Fuzzy Logic: A fuzzy logic controller (FLC) handles uncertainty and imprecision in systems, mimicking human decision-making processes, making it useful in environments where exact values cannot be determined.
- Impact of Intelligent Control: The implementation of intelligent control systems in manufacturing results in increased automation, efficiency, and the ability to respond to real-time disruptions. These systems lead to greater autonomy, flexibility, and resilience in factory operations, particularly in dynamic and unpredictable environments.

Exercises

P3.1 What is an intelligent control system?

P3.2 Name two types of intelligent control techniques.

P3.3 What is the primary function of a fuzzy logic controller (FLC)?

P3.4 In decision trees, what does each node represent?

P3.5 What is the role of evolutionary algorithms in intelligent control systems?

P3.6 How does a neural network differ from traditional control methods?

P3.7 What is meant by the term "perception layer" in intelligent control systems?

P3.8 What is the advantage of using intelligent control in manufacturing systems?

P3.9 Give an example of a real-world application of decision trees in manufacturing.

P3.10 What is the function of the "execution layer" in intelligent control architecture?

P3.11 Explain how a fuzzy logic system works in controlling the speed of a motor.

P3.12 Describe the process of training a neural network for fault detection in machinery.

P3.13 What is the importance of the "knowledge layer" in intelligent control systems, and how does it contribute to decision-making?

P3.14 How does an evolutionary algorithm optimize production scheduling in a factory?

P3.15 Discuss the advantages and limitations of using decision trees in IM systems.

4

Production Planning and Control

4.1 Introduction

Production planning and control (PPC) is concerned with the logistical problems encountered in manufacturing. It deals with ordering the materials and obtaining the resources required to produce the product in sufficient quantities to satisfy demand. Questions such as what products are to be produced, how many products are to be produced, when to produce the products, and when to obtain the raw materials, spare parts, and resources to produce the parts are answered using information extracted from vast amounts of data being generated from the actual production, functional requirements, and market assessments.

In contrast, as discussed in Chapter 2, process planning is concerned with technical details and the engineering and technological issues of how to make the product and its parts. Questions such as what types of equipment and tooling are required to fabricate the parts and assemble the product are related to process planning. Therefore, differentiating between process and production planning is essential.

Moreover, in PPC, planning and control should function as integrated activities. Creating a production plan alone is not enough if the factory resources are not managed to fulfill that plan. Similarly, controlling production efforts without a predefined plan to assess progress is ineffective. Therefore, planning and control are essential and must work in coordination with each other and other key functions within the manufacturing environment, such as process planning, concurrent engineering, and advanced manufacturing strategies (see Figure 4.1). The difference between production planning and production control is listed in Table 4.1.[1]

4.2 Production Planning

The production planning process is shown in Figure 4.2. It is divided into three stages, i.e., long-range, medium-range, and short-range. The long-range stage includes the process planning stage (see Chapter 2) and strategic capacity planning, such as allocating resources toward acquiring additional

DOI: 10.1201/9781003485186-5

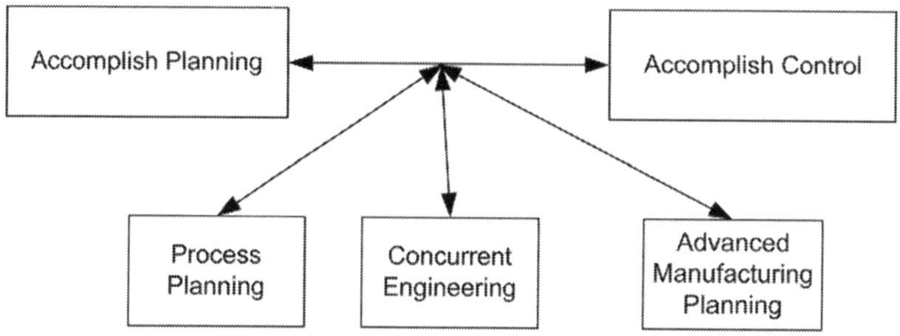

FIGURE 4.1
Integration of planning and control functions in production systems.

TABLE 4.1
Production Planning vs Production Control

Attribute	Production Planning	Production Control
Meaning	• Concerning the determination, acquisition, and arrangement of all facilities necessary for future operations • Fixes goals of production and estimates resources required to achieve the goal • Forecasts individual steps in the production process • Helps to achieve production goals effectively, promptly and economically	• Involves the coordination and integration of the factors of production for optimum efficiency • Directs, coordinates, and controls production • Implements the production plan • Done after production planning • Helps to have maximum production at minimum cost • Helps to have timely delivery of goods
Goals	Fixes the goals for production	Achieves the goals set by production planning
Consists of	• Deciding which products to make, in what quantities, and when they should be completed • Scheduling the delivery and/or production of the parts and products • Planning the manpower and equipment resources needed to accomplish the production plan	Determining whether the necessary resources to implement the production plan have been provided, and if not, attempting to take corrective action to address the deficiencies
Activities	• *Aggregate Production Planning (APP)* Determining the production output levels for the company's main product lines. This planning must be aligned and coordinated with key departments such as product design, manufacturing, marketing, and sales to ensure overall consistency and effectiveness	• *Shop Floor Control* Evaluate the real-time progress and current status of production orders on the factory floor against the planned targets outlined in the MPS and MRP schedule

(Continued)

TABLE 4.1 (*Continued*)

Production Planning vs Production Control

Attribute	Production Planning	Production Control
	• *Master Production Schedule* The APP needs to be translated into a master production schedule (MPS), which outlines the specific quantities of each individual model to be produced within every product line • *Material Requirements Planning (MRP)* Translates the MPS of end products into a detailed schedule for the raw materials and parts used in those end products • *Capacity Planning* Determines the labor and equipment resources needed to achieve the master schedule	• *Inventory Control* Encompasses a range of techniques used to manage a company's inventory effectively. A key tool in inventory control is the Economic Order Quantity (EOQ) formula, which helps determine the optimal order size to minimize costs

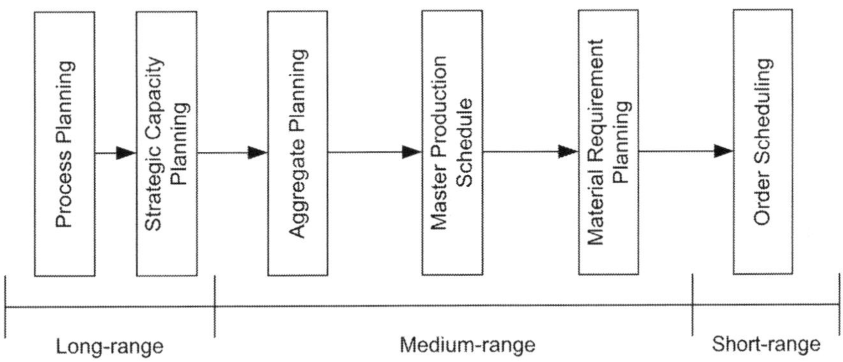

FIGURE 4.2
Stages in production planning.

machines or upgrading to more efficient ones to meet anticipated increases in production demand. Aggregate planning, master production schedule (MPS), and material requirement planning (MRP) are part of the medium range, while order scheduling constitutes the short-range stage.

4.2.1 Aggregate Production Planning (APP) and Master Production Schedule (MPS)

Aggregate production planning (APP) is a high-level strategic activity that outlines the desired output levels for the company's major product lines. Close coordination with the sales and marketing departments is required to align production capabilities and market demand. Since the plan often includes products already in production, it must account for both current and anticipated inventory levels. Moreover, planning for future inventory, especially considering new product introductions, necessitates the reconciliation of marketing strategies and promotional activities with the organization's overall capacity and resource constraints.

The production quantities outlined in the APP for major product lines must be translated into a detailed schedule known as the MPS. The MPS specifies the individual products to be manufactured, their required completion dates, and the exact quantities in which they need to be produced. It is essential to understand that MPS should be grounded in a reliable forecast of customer demand and a practical evaluation of the company's available production capacity to ensure its feasibility and effectiveness. Tables 4.2 and 4.3 show an MPS derived from a hypothetical APP.

Since MPS is considered a "medium-range plan," the adjustments are generally not permitted within a six-week horizon due to the challenges and complexities of altering established production schedules. Moreover, MPS handles the following aspects:

- Lead times to order raw materials and components
- Total parts produced in the factory
- Assembly of the end products.

TABLE 4.2

Hypothetical Aggregate Production Plan

	Week								
Product Line	1	2	3	4	5	6	7	8	9
X Model line	100	100	125	125	150	150	200	200	100
Y Model line	80	80	95	100	125	–	–	–	85
Z Model line	–	–	–	75	85	100	100	90	90

TABLE 4.3

Master Production Schedule for Table 4.2

	Week								
Product Line Models	1	2	3	4	5	6	7	8	9
Model X2	50	50	80	80	75	75	100	100	50
Model X3	50	50	45	45	75	75	100	100	50
Model Y1	30	30	40	50	70	–	–	–	–
Model Y2	50	50	55	50	55	–	–	–	–
Model Z5	–	–	–	75	85	100	100	90	90

Also, MPS is divided into three categories: firm customer orders, forecasted demand, and spare parts. The proportions of each product category in the MPS differ by company, and some may omit one or more categories. Companies assembling products typically manage all three: customer orders with promised delivery dates, forecasted demand based on past trends and sales estimates (often the largest share), and repair parts for service or direct shipment. Some companies exclude repair parts from the MPS as they are not final products.

4.2.2 Material Requirement Planning (MRP)

MRP is a structured computational approach used in manufacturing (in medium-range) to convert the MPS for finished goods into a comprehensive, time-phased schedule for the procurement of raw materials, components, and subassemblies. It plays a central role in coordinating production activities and ensuring that all necessary materials are available precisely when needed to meet production targets.

MRP is specifically designed to handle dependent demand items, i.e., those whose demand is directly determined by the demand for another item, typically a finished product. Examples of dependent demand items include raw materials, component parts, and subassemblies. Since their requirement is linked to the production of final goods, these items are not forecasted but are calculated based on the actual production schedule. In contrast, independent demand items, such as final products or spare parts, are generally forecasted and unrelated to the demand for other items.

To illustrate, while the monthly demand for a specific car model must be forecasted, once the production quantity is known, the number of components required, such as five tires per car, including a spare, can be precisely calculated.

MRP systems generate detailed schedules that specify the following:

- The quantity of each item required.
- The time at which each item must be ordered.
- The time at which each item must be delivered.

By doing so, MRP ensures that materials are available for production without overstocking or last-minute procurement inefficiencies. It is, therefore, widely regarded as an effective method of inventory control, helping organizations minimize unnecessary inventory investment, optimize storage space, and maintain smooth production flows. MRP supports lean and responsive manufacturing operations by properly managing raw materials, work-in-process (WIP), and components.

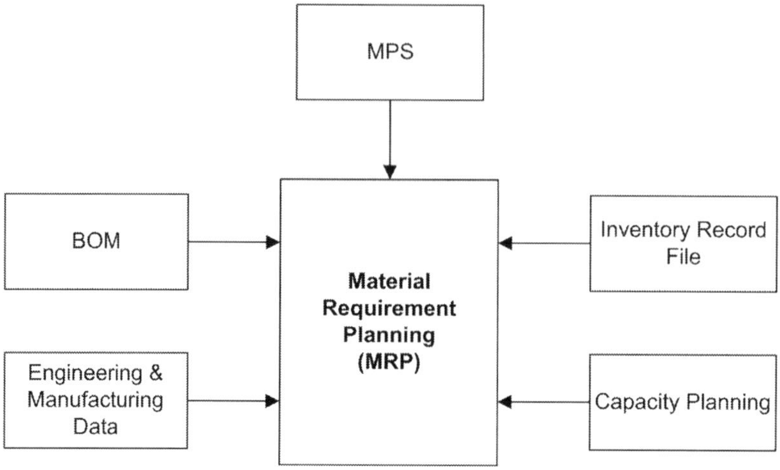

FIGURE 4.3
Inputs of MRP.

4.2.2.1 Inputs of MRP System

The inputs of the MRP system are shown in Figure 4.3 and listed below:

i. MPS.
ii. Bill of Material (BOM): A comprehensive list of all the raw materials, components, assemblies, and parts, along with their quantities, needed to manufacture a product, essentially acting as a recipe or blueprint for production.
iii. Engineering and Manufacturing Data: coming from the process planning stage (see Chapter 2).
iv. The Inventory Record File (also known as "Item master file"): contains three critical elements:
   - Item Master Data: item identification such as part number, order quantities and lead times.
   - Inventory Status: gross requirements for the items, including scheduled receipts, on-hand status, and planned order releases.
   - Subsidiary Data: purchase orders (POs), scrap or rejects, and any changes in process planning data.
v. Capacity Planning: short-term, such as hiring temporary workers to compensate for an increase in product demand, increase or decrease in the number of work shifts, etc.

4.2.2.2 How Does MRP Work?

MRP functions majorly by using data from the MPS, the BOM, and the inventory record file (as shown in Figure 4.3). It determines the quantity and timing of each required component and raw material by breaking down, or "exploding," the final product requirements into their individual parts across multiple levels of the product structure. A sample product structure is shown in Figure 4.4.

The finished product, P1, is composed of two subassemblies: S1 and S2. S1 is made up of components C1–C4, which further require raw materials RM1–RM4 in varying quantities. Similarly, S2 consists of components C5–C8, each linked to raw materials RM5–RM8. The numbers in parentheses indicate the required quantities of each part or material. This hierarchical breakdown helps MRP systems accurately plan procurement and production activities across all levels.

Example 4.1

Consider the product structure shown in Figure 4.4. Assume that 100 units of P1 are to be made. Find the quantities for subassemblies, components, and raw materials.

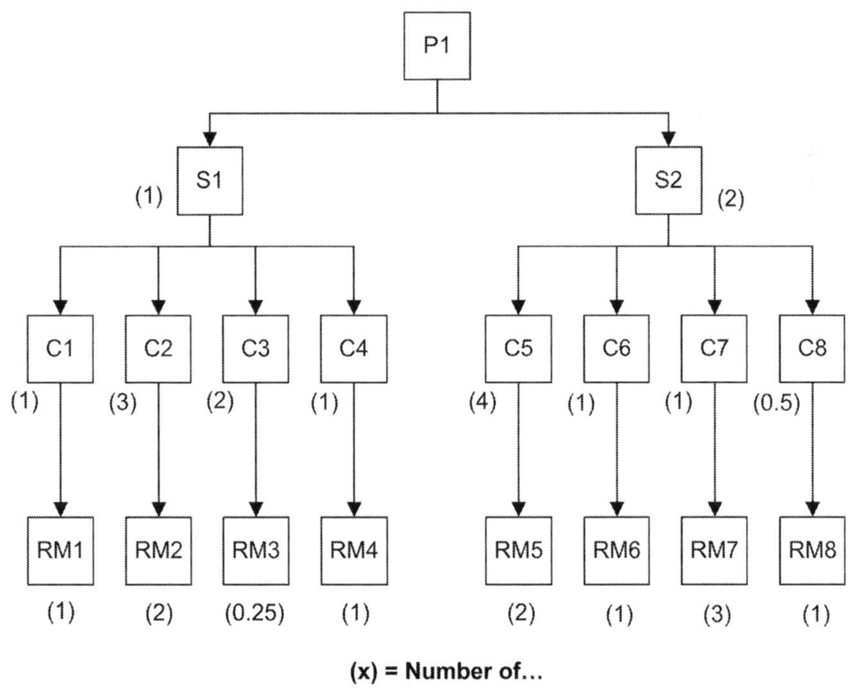

FIGURE 4.4
A product structure in MRP.

Solution:

i. At the sub-assembly level, 100 units of P1, 100 units of S1, and 200 units of S2 are required.
ii. At the component level, 100 units of S1 require 100 units of C1, 300 units of C2, 200 units of C3, and 100 units of C4. Similarly, 200 units of S2 require 800 units of C5, 200 units of C6, 200 units of C7, and 100 units of C8.
iii. At the raw material level, 100 units of C1 require 100 units of RM1, 300 units of C2 require 600 units of RM2, 200 units of C3 require 50 units of RM3, 100 units of C4 require 100 units of RM4, 800 units of C5 require 1,600 units of RM5, 200 units of C6 require 200 units of RM6, 200 units of C7 require 600 units of RM7, and 100 units of C8 require 100 units of RM8.

Example 4.1 did not consider items already available in inventory or those expected to arrive from existing POs. To address this, a process called "netting" is used to adjust the calculated material quantities. Against netting, the net requirements are determined by subtracting on-hand inventory and outstanding order quantities from the gross requirements, ensuring that only the necessary additional amounts are planned for procurement or production, as shown in Table 4.4. Therefore, whenever the calculation of netting is performed, it is imperative to incorporate the following two elements in the MRP:

- Common-use items such as raw materials and components used on more than one product.
- Lead Time—time required to complete a job from start to finish.

Lead time can be categorized into two main types: ordering and manufacturing lead time. Ordering lead time refers to the duration from raising a purchase requisition (PR) to receiving the item from the supplier, typically through a PO. In contrast, manufacturing lead time represents the time taken to produce an item within the company's facility, from the PO's release to the final completion, assuming all necessary raw materials are already available.

TABLE 4.4

Netting in MRP

Period		1	2	3	4	5
Item: **Raw Material RM3**						
Gross requirements						
Scheduled receipts				30		
On hand	20	20	20	50		
Net requirements						
Planned order releases						

Conclusively, the MPS of end products must be translated into 'time-phased requirements' for components and materials by factoring in the ordering and manufacturing lead times.

4.2.2.3 MRP Time-Phased Quantity Requirements

The time-phased quantity requirements are explained with the help of Example 4.2, as shown below.

Example 4.2

The product structures for products Y1 and Z5 are shown below in Figures 4.5 and 4.6, respectively. The inventory status and lead times are shown in Table 4.5. Complete the time-phased requirements for Y1, Z5, S1, S3, C4, and RM3 based on the requirements of Y1 and Z5 given in the MPS (see Table 4.3).

Solution:

The time-phased requirements are shown in Table 4.6.

Since quantities are in whole numbers, planned order releases for RM3 are converted to 18 and 23, for Periods 1 and 2, respectively.

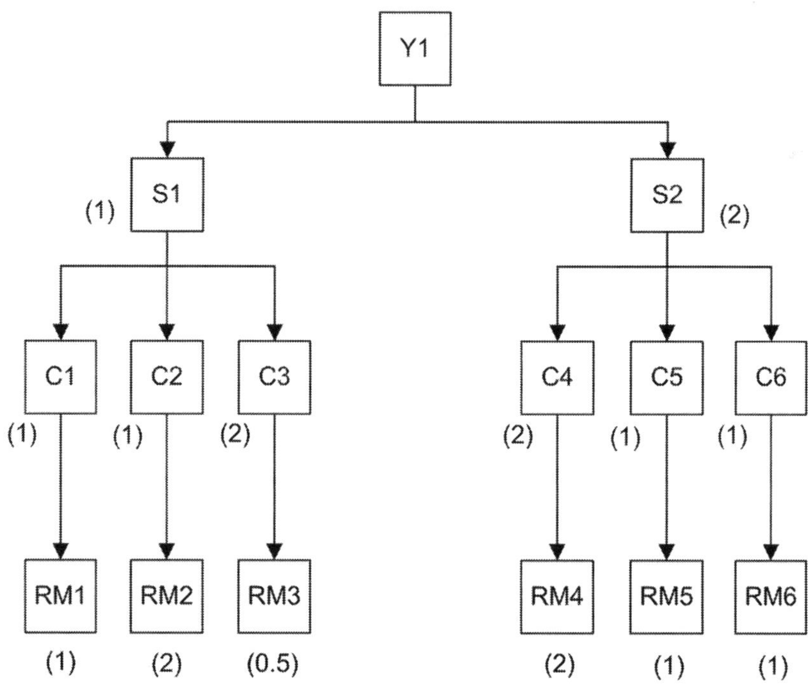

FIGURE 4.5
Product structure for Y1 (Example 4.2).

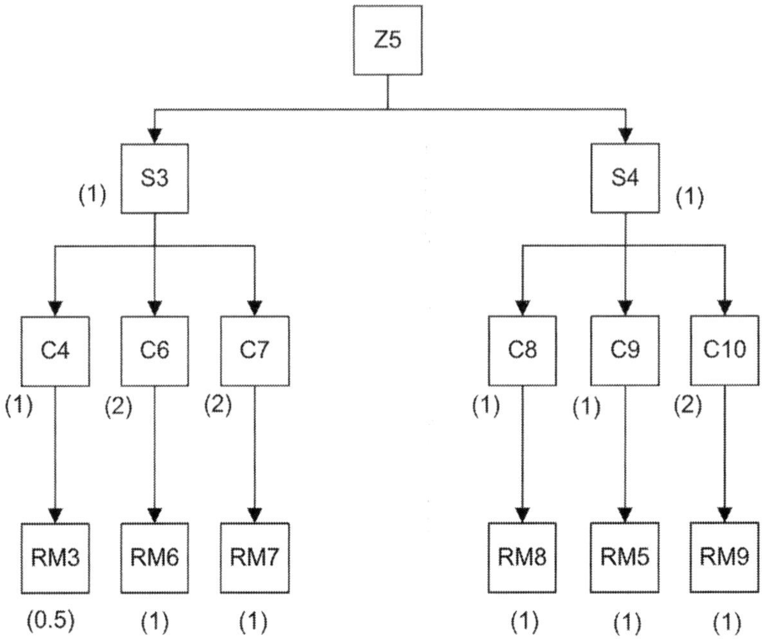

FIGURE 4.6
Product structure for Z5 (Example 4.2).

TABLE 4.5
Inventory Status and Lead Times for Example 4.2

Item	Lead Time	Inventory
Y1	Assembly lead time = 0 week	No inventory on hand or on order
Z5	Assembly lead time = 1 week	No inventory on hand or on order
S1	Assembly lead time = 0 week	No inventory on hand or on order
S3	Assembly lead time = 1 week	No inventory on hand or on order
C4	Manufacturing lead time = 1 week	No inventory on hand or on order
RM3	Ordering lead time = 0 week	As shown in Table 4.4

TABLE 4.6
MRP Time-Phased Requirements for Example 4.2

Period		1	2	3	4	5	6	7	8	9
Item: Product Y1										
Gross requirements		30	30	40	50	70	–	–	–	–
Scheduled receipts										
On hand	0									
Net requirements		30	30	40	50	70				
Planned order releases		30	30	40	50	70				

(Continued)

TABLE 4.6 (*Continued*)

MRP Time-Phased Requirements for Example 4.2

Period		1	2	3	4	5	6	7	8	9
Item: Product Z5										
Gross requirements		–	–	–	75	85	100	100	90	90
Scheduled receipts										
On hand	0									
Net requirements					75	85	100	100	90	90
Planned order releases				75	85	100	100	90	90	
Item: Sub-assembly S1										
Gross requirements		30	30	40	50	70				
Scheduled receipts										
On hand	0									
Net requirements		30	30	40	50	70				
Planned order releases		30	30	40	50	70				
Item: Sub-assembly S3										
Gross requirements				75	85	100	100	90	90	
Scheduled receipts										
On hand	0									
Net requirements				75	85	100	100	90	90	
Planned order releases			75	85	100	100	90	90		
Item: Component C4										
Gross requirements			75	85	100	100	90	90		
Scheduled receipts										
On hand	0									
Net requirements			75	85	100	100	90	90		
Planned order releases		75	85	100	100	90	90			
Item: Raw Material RM3										
Gross requirements		37.5	42.5	50	50	45	45			
Scheduled receipts				30						
On hand	20	20	20	50						
Net requirements		17.5	22.5	0	50	45	45			
Planned order releases		18	23	0	50	45	45			

4.2.2.4 Outputs and Advantages of MRP

MRP provides essential data and actions that support efficient production control and inventory management. By aligning material availability with the production schedule, MRP ensures that manufacturing runs smoothly, cost-effectively, and on time.

Primary Outputs of MRP:

i. Planned Order Releases – These are the core instructions generated by MRP, indicating the timing and quantities for initiating purchases or internal production.
    - POs are issued to external suppliers for the procurement of raw materials or components, detailing quantities and expected delivery dates.
    - Work Orders authorize the internal manufacturing of parts, assemblies, or finished goods, specifying when production should begin and how many units are required.

ii. Advance Order Release Schedules: Provide a time-phased view of future order releases to help plan resources and capacity.

iii. Rescheduling Alerts: Suggest modifications to existing orders when due dates must be updated because of changes in demand, supply delays, or capacity adjustments.

iv. Cancellation Notifications: Indicate that certain planned or released orders are no longer required, often due to revisions in the production plan or inventory status.

v. Inventory Reports: Show the current stock levels of materials and components, highlighting shortages or surplus situations.

vi. Operational Performance Reports: Deliver insights into costs, resource utilization, and efficiency by comparing actual results with planned expectations.

vii. Exception Reports: Highlight anomalies in the system such as overdue orders, scrap rates, or deviations from production targets.

viii. Inventory Projections: Estimate future stock positions to aid in strategic procurement and production decisions.

Key Benefits of MRP:

i. Lower Inventory Levels: MRP supports just-in-time procurement by ensuring materials are ordered only when needed, reducing the costs of holding excess stock.

ii. Faster Response to Market Changes: The system can quickly adjust order quantities and timing in response to fluctuations in customer demand or supply chain disruptions.

iii. Minimized Setup and Changeover Costs: Efficient planning reduces the frequency of machine changeovers, leading to cost savings and time efficiency.

iv. Enhanced Equipment Utilization: MRP helps align production with available capacity, increasing the effective use of machinery and labor.

v. Improved Planning Flexibility: Organizations can adapt more readily to changes in the MPS with updated order recommendations.
vi. Assistance in MPS Creation: MRP provides data-driven support for building or refining the Master Production Schedule by supplying detailed insights on timing, quantities, and material needs.

4.2.3 Capacity Planning

Capacity planning is the process of aligning a company's production capabilities with demand, ensuring that both short-term and long-term needs are met efficiently. It involves determining the labor and equipment resources necessary to fulfill the current MPS and preparing for future production demands.

Short-term capacity planning focuses on making quick, flexible adjustments to handle fluctuations in production demand without committing to permanent changes. These tactical and usually reversible adjustments enable firms to respond effectively to immediate operational needs. Common strategies include altering employment levels, hiring temporary workers, or modifying work shifts to scale labor input up or down as required. Companies may also adjust labor hours through overtime or reduced scheduling. Additionally, inventory stockpiling during low-demand periods helps stabilize production, while managing order backlogs allows for delayed deliveries when capacity is tight. Subcontracting is another helpful approach, enabling firms to outsource work to external vendors during peak periods or when internal resources are stretched.

Long-term capacity planning, in contrast, involves strategic decisions that shape the future production capabilities of an organization. These adjustments require significant investment and aim to meet projected demand growth over an extended period. Long-term strategies include acquiring new equipment to boost efficiency, building new plants to expand capacity, or purchasing existing facilities for quicker expansion. Companies may also acquire other firms to increase output and gain competitive advantages through economies of scale and market consolidation. Conversely, closing underutilized or obsolete plants helps reduce operational costs and optimize resource allocation. These decisions are critical in ensuring the firm remains competitive and meets future market demands.

4.3 Production Control

4.3.1 Shop Floor Control

Shop floor control (SFC) refers to the set of practices and systems used to manage, track, and optimize production activities on the factory floor. It ensures that manufacturing orders are effectively released, scheduled, and monitored to meet production targets. SFC typically operates through three sequential phases: order release, order scheduling, and order progress.

i. The necessary documentation to initiate POs is generated and processed in the order release phase. This documentation, collectively referred to as a *shop packet*, includes items such as route sheets (or process plans), material requisitions for drawing materials from inventory, job cards for tracking labor input and order status, move tickets for internal material handling, and a parts list for assembly jobs. The two primary inputs for order release are (1) output from the MRP system, which provides work orders with scheduling information, and (2) engineering and manufacturing databases that define product structure and routing sequences.
ii. The order scheduling phase assigns production orders to various work centers (WCs) in the plant using dispatch lists. This phase addresses two primary scheduling issues: machine loading (determining where jobs will be processed) and job sequencing (defining the order in which jobs are handled at each WC). Scheduling decisions are guided by *priority control* systems and follow various dispatching rules such as first-come-first-served, earliest due date, shortest processing time, least slack time (based on the time difference between remaining processing time and due date), and critical ratio (the ratio of time left until due date to remaining process time).
iii. The order progress phase involves monitoring the status of orders across the shop floor, including WIP and overall production performance. This is typically reported through several types of documents: work order status reports (detailing order location, processing progress, and schedule adherence), progress reports (summarizing performance over a time period), and exception reports (highlighting deviations from production schedules).

To support these SFC activities, MES play a vital role. MES integrates real-time data and provides functionalities such as generating process instructions, monitoring machine status, tracking labor activity, and controlling inventory. MES also links with other enterprise functions like quality control, maintenance, and product design, facilitating informed decision-making across departments.

Additionally, Factory Data Collection (FDC) systems are employed to gather real-time production data using a combination of manual methods (e.g., job travelers and employee time sheets) and automated tools [e.g., barcodes, radio frequency identification (RFID), and automatic identification and data capture (AIDC) technologies]. FDC systems collect data on completed units, scrapped parts, completed operations, labor time spent, and machine downtime. Advanced configurations include centralized terminals or workstation-specific input points. Through standards like *MTConnect*, data from machines can be analyzed to determine performance metrics such as overall equipment effectiveness (OEE), enhancing visibility and operational efficiency on the shop floor.

4.3.2 Inventory Control

Inventory control involves finding the right balance between two often competing objectives: minimizing inventory holding costs and maximizing customer satisfaction. On the one hand, reducing inventory levels to the bare minimum (sometimes even targeting zero inventory) helps lower costs associated with storage, capital investment, and the risk of spoilage. These expenses are generally known as *carrying costs* or *holding costs*. On the other hand, maintaining high inventory levels ensures customer orders can be fulfilled promptly, enhancing service levels and customer satisfaction.

However, keeping large inventories increases operational costs, while too little inventory can result in stockouts, potentially leading to lost sales and dissatisfied customers. The key lies in striking a balance between minimizing the costs of holding stock and avoiding the costs associated with running out of inventory. This trade-off is critical for efficient production and supply chain management.

4.3.2.1 Economic Order Quantity

Inventory can exist in various forms, including raw materials, purchased parts, WIP, and finished goods. Managing these inventory types effectively ensures smooth production flow while maintaining cost efficiency and service reliability. One of the important tools for inventory control is the *economic order quantity (EOQ)* formula.

The EOQ model helps determine the most cost-effective quantity of an item to purchase or produce. This method assumes a stable demand rate and that production capacity exceeds the consumption rate, an approach typically used in make-to-stock (MTS) systems. EOQ aims to minimize the total cost of ordering and holding inventory by optimizing order size.

In cases where demand remains relatively constant, especially for items with dependent demand, inventory is consumed steadily over time. When the stock reaches a certain threshold, it is replenished rapidly to a predefined maximum quantity, which is dictated by the order size. This cycle of gradual depletion followed by sudden restocking gives inventory levels a saw-tooth-shaped pattern when plotted over time, as shown in Figure 4.7. The visual pattern reflects how inventory is managed: dropping steadily as items are used and rising sharply when a new batch is received.

Figure 4.7 also shows the importance of reorder point (ROP) systems, which help determine the appropriate moment to replenish inventory.

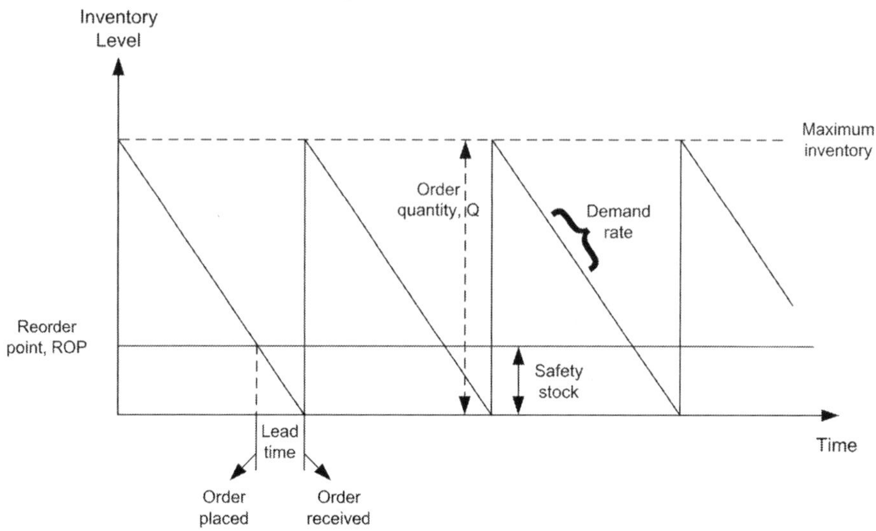

FIGURE 4.7
Inventory level over time for MTS situations.

Modern inventory control systems often automate ROP tracking, enabling timely and efficient restocking decisions.

In real-world scenarios, demand can fluctuate between placing and receiving an order, which introduces uncertainty into the process. To manage this, the ROP is set at a quantity that helps reduce the risk of stockouts during the lead time. This ensures there is enough inventory on hand until the new shipment arrives. The reorder point is typically calculated using Eq. 4.1:

$$\text{ROP} = \text{Demand during lead time} + \text{Safety stock} \tag{4.1}$$

Here, demand during lead time is the number of days between placing a PO with a supplier for a product and receiving the product. Safety stock is the extra 'just-in-case' inventory kept on hand to anticipate variability in demand or supply.

The demand during lead time and safety stock are calculated using Eqs. 4.2 and 4.3, respectively.

$$\text{Demand during lead time} = \text{Lead time}\left(\text{in days}\right) \times \text{average daily sales} \tag{4.2}$$

$$\begin{aligned}\text{Safety stock} = &\left(\text{Maximum daily orders} \times \text{Maximum lead time}\right) - \\ &\left(\text{Average daily orders} \times \text{Average lead time}\right)\end{aligned} \tag{4.3}$$

The EOQ formula is given by Eq. 4.4:

$$\text{TIC} = \frac{C_h Q}{2} + \frac{C_{\text{su}} D_a}{Q} \tag{4.4}$$

where

- TIC (Currency²/year) = total annual inventory cost (holding cost + ordering cost)
- Q (piece/order) = order quantity
- C_h (Currency/piece/year) = carrying or holding cost
- C_{su} (Currency/setup or Currency/order) = setup cost and/or ordering cost for an order
- D_a (piece/year) = annual demand for the item
- D_a / Q = Number of orders or batches produced per year (Number of setups per year).

Since C_h is the carrying or holding cost, it represents the annual expense of storing a single unit of inventory. This cost comprises two main components: investment cost and storage cost.

- Investment cost is calculated as the product of the company's annual interest rate (i) and the value (cost) of the inventory.
- Storage cost accounts for the item's physical space and handling requirements, often expressed as a percentage (s) of the item's value.

These two items, i and s, combine to form the annual holding cost rate given by the formula $h = i + s$. Thus, the holding cost, C_h, is provided by Eq. 4.5:

$$C_h = hC_{\text{pc}} \tag{4.5}$$

where C_{pc} is the item's unit cost (Currency/piece).

In Eq. 4.4, C_{su} is the setup cost or ordering cost for an order. This cost is impacted by the time the production equipment is idle during changeover time between batches and the cost associated with the labor that does the setup changes. Therefore, C_{su} can be given by Eq. 4.6:

$$C_{\text{su}} = T_{\text{su}} C_{\text{dt}} \tag{4.6}$$

where T_{su} is the setup or changeover time between batches (in time/setup or time/order) and C_{dt} is the cost rate of machine downtime during the changeover (in Currency/hour). The setup cost should also include the internal costs of placing the order with the vendor.

It is important to note here that Eq. 4.4 excludes the actual annual cost of the part's production. If the cost associated with the annual demand is added, Eq. 4.4 is converted as shown in Eq. 4.7 shows the annual total cost:

$$\mathrm{TC} = D_a C_{\mathrm{pc}} + \frac{C_h Q}{2} + \frac{C_{\mathrm{su}} D_a}{Q} \tag{4.7}$$

where $D_a C_{\mathrm{pc}}$ is the product of annual demand (piece/year) and cost per item (Currency/piece).

For Eq. 4.4 or Eq. 4.7, take the derivative, put it equal to zero, and solve for Q. This results in the EOQ formula as given in Eq. 4.8. EOQ computes the number of parts to be produced per batch and has units of piece/batch or piece/order:

$$Q = \mathrm{EOQ} = \sqrt{\frac{2 D_a C_{\mathrm{su}}}{C_h}} \tag{4.8}$$

Example 4.3

The annual demand for a certain item MTS is 10,000 pieces/year. One unit of the item costs USD 200, and the holding cost rate is 15% per year. The setup time to produce a batch is 6 hours. The cost of equipment downtime plus labor is USD 200 per hour. Determine the economic batch quantity and the total inventory cost.

Solution:

Setup cost is given by:

$$C_{\mathrm{su}} = T_{\mathrm{su}} C_{\mathrm{dt}} = 6 \times 200 = \mathrm{USD}\ 1{,}200$$

Holding cost per unit = 0.15 × 200 = USD 30. Now, EOQ is

$$\mathrm{EOQ} = \sqrt{\frac{2(10{,}000)(1{,}200)}{30}} = 894.4 \approx 894 \text{ units}$$

Now, TIC can be calculated as:

$$\mathrm{TIC} = \frac{(30)(894)}{2} + \frac{(1{,}200)(10{,}000)}{894} = 13{,}410 + 13{,}423 = \mathrm{USD}\ 26{,}833$$

Including the annual production costs in the annual total,

$$\mathrm{TC} = (10{,}000)(200) + 26{,}833 = \mathrm{USD}\ 2{,}026{,}833$$

4.3.3 Manufacturing Resource Planning (MRP-II)

Manufacturing Resource Planning (MRP-II) is an advanced, computer-integrated system that emerged in the 1980s as an evolution of the original MRP system. Unlike its predecessor, which primarily focused on materials planning, MRP-II expands its scope to include the planning, scheduling, and control of materials, labor, machinery, and supporting resources.

The system is structured as a closed-loop framework that ensures continuous feedback and coordination among the various functional areas involved in manufacturing. It aims to align production activities with the MPS through integrated modules such as MRP, capacity planning, and SFC. By providing a unified platform, MRP-II supports the synchronization of business operations, enhances decision-making, and improves overall manufacturing efficiency.

4.3.4 Enterprise Resource Planning (ERP)

ERP is an integrated information system that consolidates all essential business functions such as sales, marketing, procurement, design, production, logistics, finance, and human resources, into a single, centralized database. Each functional module within an ERP system is built on industry best practices, ensuring process consistency and operational efficiency across the organization.

ERP systems typically operate as client-server architectures, allowing users to access the system from their individual workstations across the enterprise. Unlike plant-specific systems like MRP, ERP works company-wide, ensuring that all departments operate from the same data foundation. While access is granted organization-wide, data visibility is role-based to maintain security and relevance.

A key benefit of ERP is using a unified database, which reduces data redundancy, eliminates inconsistencies, and enhances interdepartmental communication. This centralization ensures that accurate, real-time information is available to support strategic decision-making.

As ERP systems have matured, new generations have emerged. ERP II extends access to external stakeholders, such as vendors, enabling collaborative planning and integration. ERP III takes this further by involving customers in the value chain, allowing real-time communication and transparency across all nodes of the enterprise network. These advances reflect ERP's evolution from internal optimization tools to broad-based collaboration platforms.

4.4 Highlights

Highlights of the chapter on Production Planning and Control are as follows:

a. PPC integrates planning and monitoring activities to ensure timely, cost-effective manufacturing that aligns with market demand and production capabilities.
b. Production planning operates at three levels:
    - Long Term: Strategic capacity planning
    - Medium Term: APP, MPS, and MRP
    - Short Term: Order scheduling and execution.
c. APP sets high-level output goals for product lines, while the MPS translates these into detailed production schedules for individual items.
d. MRP converts the MPS into a time-phased plan for components and raw materials, focusing on dependent demand and incorporating lead times and inventory status.
e. MRP outputs include planned order releases, rescheduling notices, exception reports, and inventory forecasts, helping minimize inventory costs and avoid material shortages.
f. Capacity Planning ensures that production resources (labor and equipment) align with current and future demands.
    - Short-Term Strategies: Overtime, shift adjustments, temporary hires, subcontracting
    - Long-Term Strategies: Investment in equipment, plant expansion, acquisition, or closure.
g. SFC manages the release, scheduling, and monitoring of production orders. It involves real-time tracking supported by MES and FDC.
h. Inventory Control balances holding costs and customer service levels. Techniques include ROP systems and the EOQ model, which helps determine optimal batch sizes.
i. MRP-II extends MRP to include all manufacturing resources (labor, equipment, and materials) with feedback control.
j. ERP integrates all enterprise functions (finance, HR, sales, and manufacturing) into a unified system with real-time data access, supporting better decision-making.

Exercises

P4.1 Why should dependent demand items not be forecasted in an MRP system?

P4.2 How is the MPS different from the Aggregate Production Plan (APP)?

P4.3 What two costs are balanced in the Economic Order Quantity (EOQ) model?

P4.4 In what major way does ERP differ from MRP-II?

P4.5 **Application-based problem**

A company manufactures precision gearboxes used in automated machines. According to their Master Production Schedule (MPS), they must produce 150 units of Gearbox Model GX-200 in Week 5. Each GX-200 requires:

- 1 unit of Housing Assembly (HA)
- 2 units of Gear Set (GS)
- 4 units of Bearing (B)

Additional details:

- Current on-hand inventory: HA = 10, GS = 40, B = 100
- Scheduled receipts arriving in Week 4: HA = 50, GS = 60, B = 120
- Lead time: HA = 1 week, GS = 2 weeks, B = 1 week
- The production department can only assemble 100 gearboxes per week due to labor constraints.

Based on the MPS and available data, (a) would the company be able to meet the Week 5 requirement of 150 gearboxes? (b) If not, how should the company reschedule production or adjust capacity to meet demand? (c) How many units of Bearings (B) need to be ordered in Week 2 to meet the Week 5 schedule?

Note for Students: Use MRP logic to calculate gross and net requirements. Clearly show your working for each answer.

Notes

1 Groover, M. P. (2015). *Automation, Production Systems, and Computer-Integrated Manufacturing* (4th ed.). Princeton, NJ: Pearson.

2 Currency represents applicable monetary units like USD, EUR, PKR, etc.

5

Sensors

5.1 Introduction to Conventional Sensors

Sensors are an integral part of manufacturing processes, enabling real-time monitoring, data collection, and feedback for optimization. Conventional sensors are responsible for providing critical data regarding variables such as temperature, pressure, humidity, position, and motion. These sensors have laid the foundation for automated processes and real-time decision-making in industrial settings. In the context of IM, the role of conventional sensors is pivotal, especially in terms of their reliability, accuracy, and cost-effectiveness.

Conventional sensors are typically defined by their physical form and the type of measurement they perform. They operate based on established principles of physics and engineering and are often categorized based on their operational mechanism (e.g., mechanical, electrical, optical, or chemical sensors).

5.1.1 Types of Conventional Sensors

In this section, the most common types of conventional sensors found in manufacturing and industrial applications are introduced.

a. Temperature Sensors: Temperature sensors are essential in several manufacturing applications where temperature regulation is critical. These sensors measure the temperature of a substance or environment, ensuring that it stays within specific parameters necessary for optimal operation.

   Types of temperature sensors

   - Thermocouples: Thermocouples are one of the most widely used temperature sensors due to their durability, speed, and versatility. They consist of two different metal wires joined at one end. When the junction is heated or cooled, it generates a voltage that corresponds to the temperature. Thermocouples are commonly used in extreme environments and can measure a wide range of temperatures. An example of a low-cost thermocouple[1] is shown in Figure 5.1.
   - Resistance Temperature Detectors (RTDs): RTDs operate by measuring the resistance of a material, typically platinum, which

 DOI: 10.1201/9781003485186-6

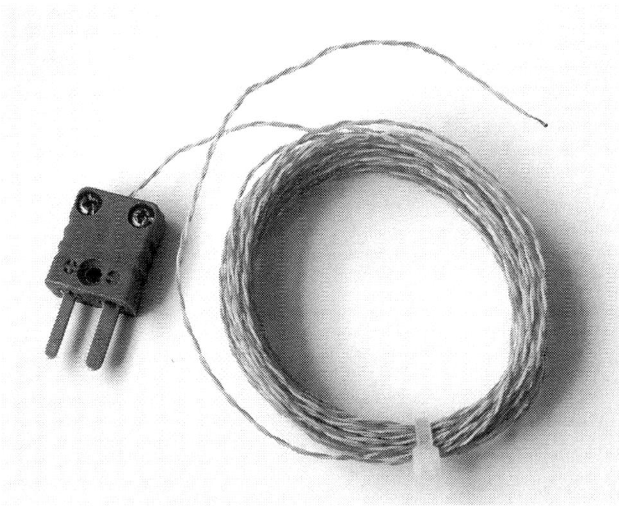

FIGURE 5.1
Example of a low-cost thermocouple.

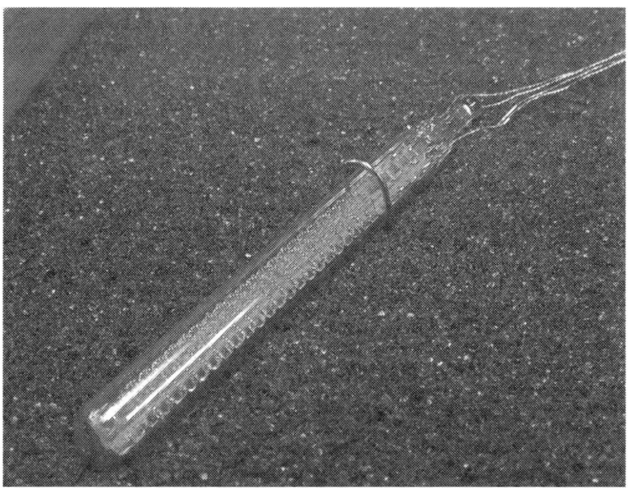

FIGURE 5.2
Example of an RTD.

increases with temperature. These sensors are highly accurate and stable, making them ideal for applications requiring precise temperature measurements, such as in the pharmaceutical industry. An example of RTD[2] is shown in Figure 5.2.

- Thermistors: Thermistors are temperature-dependent resistors that provide a more sensitive response to temperature changes than RTDs. They are typically used in applications requiring

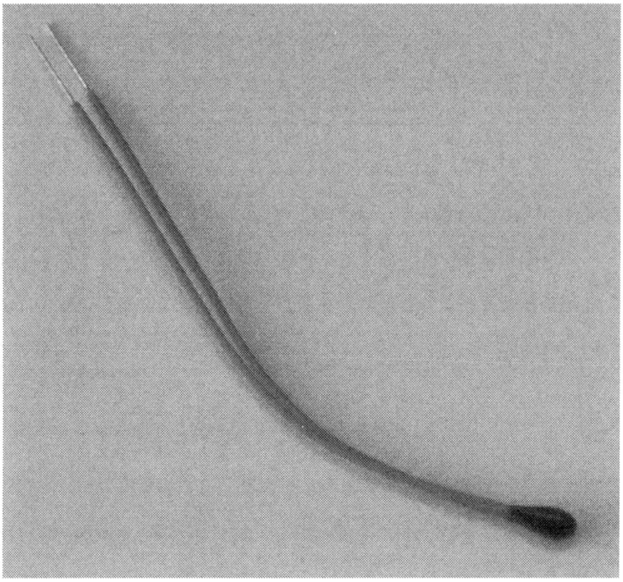

FIGURE 5.3
Example of a thermistor.

high precision, such as in consumer electronics or automotive systems. An example of a thermistor[3] is shown in Figure 5.3.

Temperature sensors ensure that systems such as ovens, furnaces, heating, ventilation, and air-conditioning (HVAC) systems, and chemical reactors maintain optimal conditions, which is critical for process quality, safety, and product consistency.

b. Pressure Sensors: Pressure sensors measure the force exerted by a fluid or gas on a surface. These sensors are commonly used in manufacturing environments to monitor and control the pressure in pipes, tanks, and various equipment. They play a crucial role in ensuring that systems operate within safe pressure limits and can also be used for process optimization.

Types of pressure sensors

- Strain Gauge Pressure Sensors: Strain gauges work by measuring the deformation of a material under pressure. When the material deforms, the strain gauge detects the change in resistance, which is proportional to the pressure. Strain gauge sensors (example shown in Figure 5.4[4]) are commonly used in hydraulic and pneumatic systems.
- Piezoelectric Pressure Sensors: These sensors utilize materials that generate an electric charge when subjected to mechanical stress. They are widely used in dynamic pressure measurements,

FIGURE 5.4
Example of a strain gauge sensor.

such as in engines or air compressors, where rapid changes in pressure occur.

- Capacitive Pressure Sensors: Capacitive sensors detect changes in capacitance when the pressure changes the distance between two electrodes. They are highly accurate and can be used for both low and high-pressure applications.
- Pressure sensors are critical in applications such as boilers, oil and gas systems, and pharmaceutical production, ensuring the safety and efficiency of these systems by maintaining correct pressure levels.

c. Proximity Sensors: Proximity sensors detect the presence or absence of an object without physical contact, making them essential for automation and safety applications. These sensors are commonly used to detect the position of objects in assembly lines or robotic systems.

Types of proximity sensors

- Inductive Proximity Sensors: These sensors detect metallic objects by creating an electromagnetic field. When a metal object enters the field, it induces a change in the sensor's current. Inductive sensors are widely used in industrial automation to detect the position of metallic components or machines in automated systems. An example of an inductive proximity sensor[5] is shown in Figure 5.5.
- Capacitive Proximity Sensors: Capacitive sensors detect objects by measuring changes in capacitance. They can detect both metallic and non-metallic objects, such as plastic or wood, making them versatile for various applications.

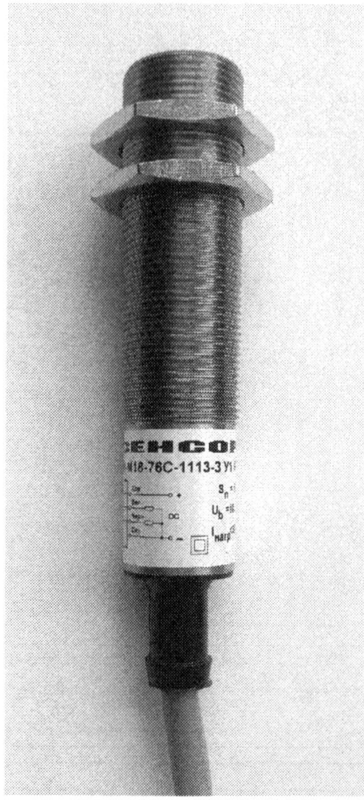

FIGURE 5.5
Example of an inductive proximity sensor.

- Ultrasonic Proximity Sensors: These sensors use sound waves to detect the presence of objects. They emit ultrasonic waves, and when these waves bounce off an object, they are detected by the sensor, allowing it to calculate the distance to the object. Ultrasonic sensors are widely used in robotics and conveyor systems for precise distance measurement.
- Photoelectric Proximity Sensors: These sensors work by emitting a light beam (usually infrared) and detecting the reflection of the light from an object. Photoelectric sensors are ideal for detecting transparent or small objects and are often used in packaging, sorting, and material handling applications.

  Proximity sensors are key components in automated assembly lines, conveyor systems, and robotic arms, enabling precise position tracking and material handling without direct contact.

d. Motion Sensors: Motion sensors detect movement or changes in position of objects in their detection field. These sensors are crucial for systems that require real-time motion tracking or are designed for security and automation purposes.

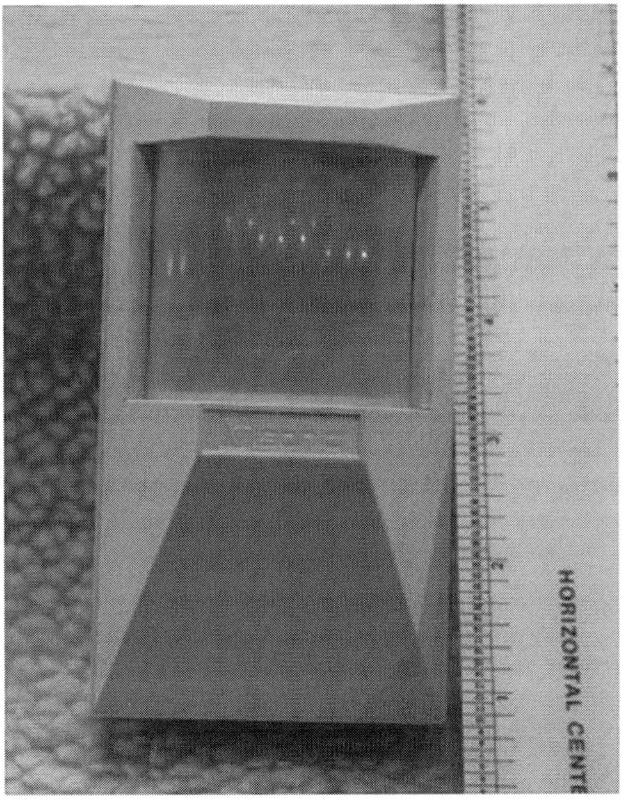

FIGURE 5.6
Example of a PIR sensor.

Types of motion sensors

- Passive Infrared (PIR) Sensors: PIR sensors detect infrared radiation emitted by living organisms or warm objects. They are often used in security systems to detect human presence and are frequently found in lighting systems, alarms, and HVAC systems to reduce energy consumption. An example of a PIR[6] sensor is shown in Figure 5.6.
- Microwave Motion Sensors: These sensors use high-frequency microwave signals to detect movement. They are capable of detecting motion through walls and other obstacles and are used in industrial automation and security systems.
- Ultrasonic Motion Sensors: Similar to ultrasonic proximity sensors, these motion sensors use sound waves to detect movement. They are often used in industrial applications to detect the movement of materials or workers, enabling automated systems to adjust accordingly.

Motion sensors play an important role in security systems, building automation, and production lines, helping to automate processes, enhance safety, and reduce energy consumption.

e. Light Sensors: Light sensors are used to detect the intensity of light in an environment. They are commonly employed in systems that require light control, such as in lighting automation, display systems, and optical measurement devices.

Types of light sensors

- Photocells (Light-Dependent Resistors – LDRs): LDRs change their resistance based on the intensity of light. They are commonly used in automatic lighting systems, where they control the lighting based on ambient light levels. An example of an LDR[7] sensor is shown in Figure 5.7.
- Photodiodes: Photodiodes generate a current when exposed to light, which is proportional to the intensity of the light. They are often used in optical systems, laser range finders, and optical communications. An example of a photodiode[8] sensor is shown in Figure 5.8.

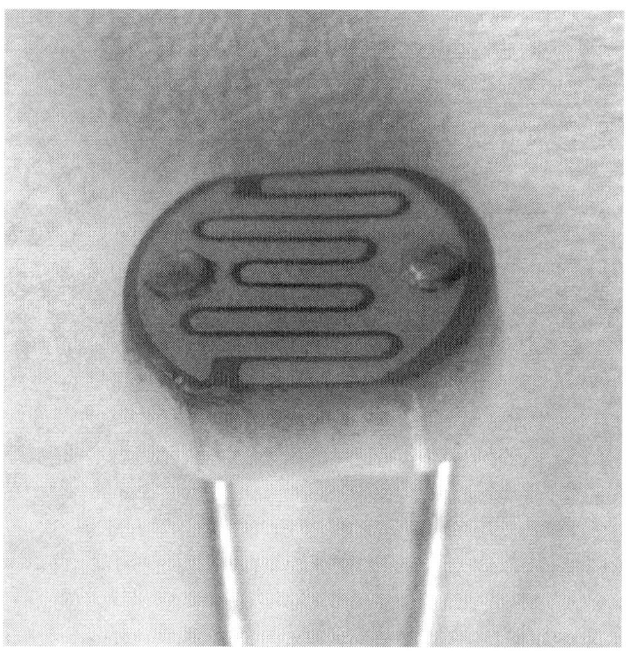

FIGURE 5.7
Example of a light-dependent resistor.

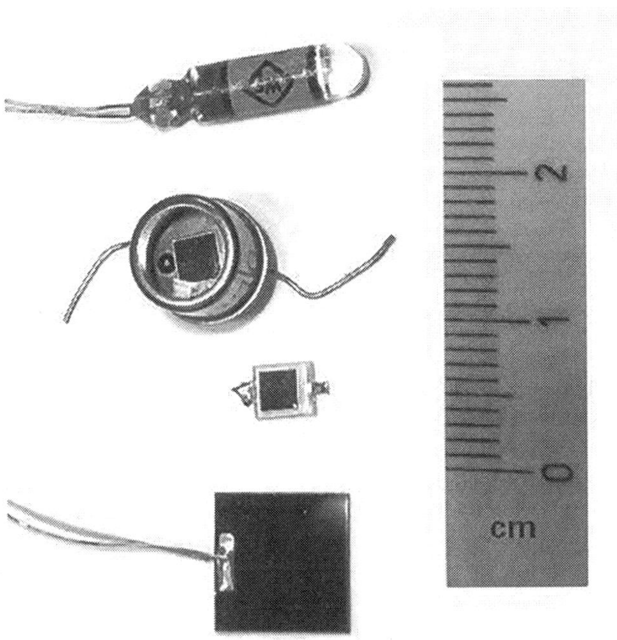

FIGURE 5.8
Photodiodes.

- Photoelectric Sensors: Photoelectric sensors can detect the presence or absence of an object by emitting a light beam and measuring the reflection or interruption of the light.

  Light sensors are widely used in automated lighting systems, display panels, and optical inspection systems in manufacturing environments.

5.1.2 Applications in Manufacturing

Below are three major areas where conventional sensors significantly contribute to manufacturing: quality control and inspection, process monitoring, and inventory management.

5.1.2.1 Quality Control and Inspection

Ensuring consistent product quality is a central goal in manufacturing. Quality control involves validating that the final product meets predefined specifications, while inspection refers to evaluating intermediate and final stages to detect defects or deviations early.

Key sensor applications in quality control:

- Dimensional Accuracy: Laser displacement sensors, optical proximity sensors, and vision systems are used to measure dimensions with micrometer precision. In industries like aerospace or automotive, where tolerances are extremely tight, these sensors help detect variations in length, width, or depth that could compromise performance or safety.
- Surface Inspection: Capacitive, ultrasonic, and optical sensors are employed to detect surface irregularities such as cracks, scratches, or corrosion. Optical sensors can scan for color consistency and pattern alignment, particularly in packaging and textile industries.
- Material Integrity: Ultrasonic and eddy current sensors can non-invasively evaluate internal defects like voids, porosity, or inclusions in materials. This is especially important in critical applications such as weld inspections, cast parts, and pressure vessels.
- Force and Torque Measurement: Strain gauge-based sensors monitor mechanical stresses applied during assembly or product usage. Over-tightening bolts or insufficient clamping forces can lead to product failure, and real-time feedback helps prevent these defects.
- Environmental Monitoring During Production: Temperature and humidity sensors are essential during sensitive processes like pharmaceutical production or semiconductor fabrication, where even minor environmental variations can compromise product quality.

By integrating these sensors into automated inspection systems, manufacturers can achieve zero-defect goals, reduce waste, and enhance compliance with stringent regulatory standards.

5.1.2.2 Process Monitoring

Process monitoring ensures that every stage of the manufacturing workflow operates within optimal parameters. It involves real-time tracking of process variables to detect deviations, prevent faults, and maintain efficiency.

Key sensor applications in process monitoring are:

- Temperature and Pressure Sensors: Critical in casting, extrusion, injection molding, and chemical processing, these sensors monitor process temperatures and pressures to ensure that materials are formed or reacted correctly. Overheating or undercooling can lead to product defects or hazardous situations.
- Vibration Sensors: Used to detect anomalies in rotating machinery, such as motors, pumps, and compressors. Early detection of

imbalances or misalignments helps avoid catastrophic failures and supports condition-based maintenance.

- Flow and Level Sensors: In fluid-based systems, sensors measure the flow rate and levels in tanks or pipes, ensuring that materials are delivered and processed at the correct volumes and rates. Inconsistent flow can result in off-spec products or process inefficiencies.
- Proximity and Motion Sensors: Deployed to monitor the presence, orientation, and movement of parts on conveyor belts, assembly robots, and CNC machines. These sensors help prevent bottlenecks and synchronize operations for higher throughput.
- Gas and Chemical Sensors: Monitor the concentration of gases (like oxygen, carbon dioxide, or volatile organics) in environments such as paint booths, cleanrooms, or chemical plants. This ensures worker safety and protects sensitive processes from contamination.

Process monitoring also supports predictive analytics, where historical sensor data is used to anticipate failures or process drift. The integration of conventional sensors with Supervisory Control and Data Acquisition (SCADA) systems and PLCs allows for dynamic control and fault mitigation in real time, driving operational excellence.

5.1.2.3 Inventory Management

Effective inventory management is a basis of lean manufacturing, ensuring the right materials are available at the right time and in the right quantities. Conventional sensors are increasingly used to automate and optimize inventory tracking, minimize stockouts, reduce overstocking, and enhance material flow.

Key sensor applications in inventory management include:

- RFID and Barcode Scanners: While not traditional analog sensors, RFID readers and barcode scanners are essential in real-time asset tracking. These sensors identify items during receiving, storage, and dispatch, ensuring accurate recordkeeping and facilitating just-in-time (JIT) inventory systems.
- Weight and Load Sensors: Load cells and strain gauges are used to measure the weight of raw materials, finished goods, or bulk items in silos and containers. This information is critical for understanding stock levels, replenishment needs, and shipping requirements.
- Level Sensors: Ultrasonic, capacitive, and radar level sensors monitor the levels of liquids, powders, or granular materials in storage bins and tanks. These sensors trigger alerts or automate restocking processes, reducing manual checks and preventing production halts due to material shortages.

- Environmental Sensors for Storage Conditions: Temperature, humidity, and light sensors ensure optimal storage conditions for sensitive items such as chemicals, food products, or pharmaceuticals. These sensors can trigger alerts or activate HVAC systems to maintain compliance and prevent spoilage.
- Proximity and Motion Sensors in Warehousing: These are used to track pallet and forklift movements, enabling warehouse management systems (WMS) to optimize layout, reduce travel times, and prevent collisions.

  Advanced inventory systems use sensor data to feed into ERP systems, ensuring real-time synchronization of supply chain activities. This leads to better demand forecasting, improved vendor coordination, and reduced working capital costs.

5.2 Introduction to Soft Sensors

In the rapidly evolving landscape of IM, where agility, real-time decision-making, and data-driven optimization are paramount, the limitations of conventional sensors have become increasingly evident. While physical (hardware-based) sensors are essential for capturing real-world measurements, they can be costly to deploy at scale, limited by environmental conditions, or unable to directly measure certain parameters. Soft sensors are a powerful and transformative class of virtual sensing technologies that leverage mathematical models, machine learning algorithms, and data fusion techniques to estimate or predict unmeasurable process variables in real time.

Soft sensors (also known as virtual sensors, inferential sensors, or software sensors) have gained widespread adoption in advanced manufacturing, particularly where physical measurements are either infeasible or inefficient. They enable manufacturers to extract greater value from existing sensor infrastructure, facilitate predictive analytics, and enable a more holistic understanding of complex manufacturing systems.

5.2.1 Principles and Functionality

Soft sensors operate by building mathematical or data-driven models that relate measurable input variables (captured via physical sensors) to non-measurable or difficult-to-measure target variables. These models may be based on first-principles physics, data analytics, or hybrid methods, and are implemented in software running on edge devices, controllers, or centralized cloud systems.

5.2.1.1 Core Principles

- Model-Based Estimation: Soft sensors use models that are either deterministic (e.g., equations derived from thermodynamics or fluid dynamics) or empirical (e.g., regression and neural networks) to estimate process parameters.
- Real-Time Operation: These models run in real time, constantly receiving live sensor inputs and computing estimations or forecasts that can be used for control, optimization, or monitoring.
- Feedback and Adaptation: Some soft sensors incorporate self-learning or adaptive capabilities, recalibrating themselves as more data becomes available or operating conditions change.

5.2.1.2 Key Functional Advantages

- Cost-Efficiency: By replacing or complementing expensive or intrusive physical sensors, soft sensors lower capital expenditure.
- Accessibility of Inaccessible Variables: Parameters like chemical composition, viscosity, or internal temperatures that are hard to measure directly can be estimated using soft sensors.
- System Understanding and Diagnosis: Soft sensors can detect hidden process anomalies or inefficiencies that physical sensors alone cannot uncover.
- Data Enrichment: They enrich sensor data with contextual or predictive insights, thereby enhancing the capabilities of supervisory systems.

  Examples include estimating the remaining useful life (RUL) of a machine, predicting product quality before the final stage, or inferring stress levels in complex mechanical systems.

5.2.2 Data Fusion and Prediction

One of the most powerful aspects of soft sensors is their ability to fuse data from multiple physical sensors, creating a synthesized view of the system's state that is both more complete and more reliable than any individual signal.

5.2.2.1 Data Fusion Techniques

- Low-Level Fusion: Combines raw data from multiple sensors before processing, often used in redundancy and noise reduction applications.

- Mid-Level Fusion: Merges features or descriptors extracted from sensor data (e.g., frequency patterns, statistical features) to form a richer representation.
- High-Level Fusion: Integrates decisions or inferences from different models to derive a final estimation or prediction.

5.2.2.2 Prediction and Machine Learning Integration

Soft sensors frequently utilize machine learning algorithms to uncover complex, nonlinear relationships between input variables and target outputs. Common techniques include:

- Support Vector Machines (SVM)
- Random Forests
- ANN
- Deep Learning (DL) models, such as Long Short-Term Memory (LSTM) for time-series forecasting.

Predictive modeling allows soft sensors not only to estimate current values but also to forecast future trends, which is essential for preventive interventions and optimizing production parameters before issues arise.

By continuously learning from historical and real-time data, these models evolve with the system, improving accuracy and robustness over time.

5.2.3 Implementation in Manufacturing

Soft sensors are increasingly being deployed across the manufacturing lifecycle, from raw material input to final product shipment. Their applications span predictive maintenance, process optimization, and inventory forecasting, all of which contribute to higher productivity, lower costs, and increased reliability.

5.2.3.1 Predictive Maintenance

Predictive maintenance involves identifying signs of equipment degradation or potential failure before they result in unplanned downtime. Soft sensors are pivotal in this approach by estimating unmeasured parameters such as:

- Internal wear levels
- Lubricant quality

- Motor winding temperatures
- Shaft misalignments.

By fusing data from vibration sensors, temperature probes, current meters, and other sources, soft sensors can predict failure modes long before they become critical. Machine learning models trained on historical failure data can trigger maintenance alerts when patterns deviate from healthy operation baselines. This allows companies to shift from scheduled to condition-based maintenance, reducing unnecessary servicing and extending equipment life.

5.2.3.2 Real-Time Process Optimization

Soft sensors enable dynamic adjustment of production parameters based on inferred process states. In high-precision or continuous manufacturing environments such as chemical, pharmaceutical, or food industries, direct measurements of key quality attributes (e.g., purity, texture, and composition) may not be available in real time.

Soft sensors fill this gap by estimating these attributes from available sensor data (e.g., temperature, flow rate, and pH), allowing for real-time feedback control. This ensures consistent product quality, reduces rework, and supports closed-loop optimization strategies.

For example, in plastic extrusion processes, soft sensors can estimate melt viscosity in real time, allowing for better control of pressure, speed, and cooling rates.

5.2.3.3 Inventory Forecasting

In modern supply chains, having accurate insights into current and future inventory levels is vital for meeting demand without overstocking. Soft sensors integrate sensor data (e.g., weight, level, and barcode scans) with ERP and MES data to build predictive inventory models. These models forecast:

- Material consumption rates
- Replenishment timing
- Potential bottlenecks or shortages.

This is particularly valuable in JIT environments, where delays can disrupt entire production cycles. Soft sensors help balance supply and demand by enabling predictive restocking, reducing idle time and capital lock-in.

Additionally, they can consider external data such as supplier lead times, historical usage trends, and even market forecasts, making inventory decisions smarter and more aligned with business goals.

5.3 Introduction to Smart Sensors

In the era of Industry 4.0, the conventional role of sensors has been transformed to a great extent. In addition to basic signal transmission, today's smart sensors function as intelligent edge devices capable of data processing, self-diagnostics, communication, and decision-making. These sensors form the foundation of CPS, enabling factories to transition from reactive workflows to proactive and adaptive manufacturing systems.

Smart sensors differ fundamentally from conventional or soft sensors by combining the advantages of both hardware and software capabilities. They typically consist of a sensing element, signal conditioning, microprocessor units, communication interfaces, and often embedded algorithms. In manufacturing contexts, they collect data, analyze it, interpret it, and trigger actions in real time.

5.3.1 Features and Capabilities

Their key features and capabilities include:

a. Integrated Processing Power: Smart sensors are equipped with microcontrollers or digital signal processors (DSPs) that allow them to preprocess raw data before transmission. This reduces the load on central systems and minimizes latency.

   Example: A vibration sensor may calculate Root Mean Square (RMS) or Fast Fourier Transform (FFT) values on-board and send only meaningful summaries to the controller.

b. Self-Calibration and Diagnostics: Modern smart sensors can self-calibrate to account for drift, aging, or environmental influences. Additionally, they often perform health checks, reporting their own status and flagging faults, thereby improving system reliability.

c. Multi-Sensing Capabilities: Some smart sensors incorporate multiple sensing modalities (e.g., temperature, pressure, and humidity in a single device), offering richer, more contextual data.

d. Edge Computing: With embedded analytics, smart sensors enable edge computing by processing data at the source. This supports faster decisions, especially in time-sensitive applications like robotics or CNC operations.

e. Plug-and-Play and Interoperability: Thanks to standard communication protocols like IO-Link, Modbus, and OPC-UA, smart sensors support easy integration, device recognition, and interoperability across diverse platforms.

f. Adaptability and Learning: Some smart sensors include machine learning capabilities, allowing them to detect new patterns over time and adapt to changing environments or workloads.

These features collectively support autonomous manufacturing environments, where machines and systems can self-organize, learn, and evolve.

5.3.2 IoT and Connectivity

The backbone of smart sensor functionality lies in their connectivity, that is, the ability to interact with machines, cloud platforms, and other sensors to enable an IoT ecosystem.

a. Communication Protocols: Smart sensors support a wide range of wired and wireless communication standards, including:
   - Ethernet/IP, PROFINET, CAN Bus (for industrial automation)
   - Wi-Fi, Bluetooth, ZigBee, LoRaWAN (for flexibility and mobility)
   - 5G/6G integration for ultra-low latency and high-bandwidth applications

b. Cloud Integration: Data collected by smart sensors can be streamed directly to cloud platforms, enabling advanced analytics, predictive maintenance algorithms, and global monitoring dashboards. Advantages include:
   - Manufacturers can access real-time key performance indicators (KPIs) from any location.
   - Multi-site operations can be benchmarked for performance optimization.

c. Cybersecurity in Sensor Networks: Smart sensors are equipped with secure boot, encrypted data channels, and device authentication protocols to protect against cyberattacks and data breaches.

d. Inter-Sensor Collaboration: Through IoT frameworks, smart sensors can form sensor networks, where devices communicate and cooperate, enhancing decision-making. For example, temperature and humidity sensors can collaboratively adjust climate control in sensitive production environments.

e. Edge-to-Cloud Continuum: Smart sensors function as key nodes in an edge-to-cloud continuum. They filter and preprocess data at the edge, send only high-value data to the cloud, and receive insights or control parameters back for execution. This connectivity empowers real-time intelligence, adaptive control systems, and full supply chain visibility in intelligent manufacturing environments.

5.3.3 Role in Intelligent Manufacturing

In IM, smart sensors support closed-loop control systems, autonomous decision-making, and advanced digital twins. Their role spans the entire manufacturing lifecycle, from design to delivery, enhancing productivity, efficiency, safety, and quality.

5.3.3.1 Collaborative Manufacturing

Smart sensors facilitate collaborative manufacturing by enabling seamless coordination between humans, machines, and digital systems. This is especially vital in environments that include collaborative robots (cobots), modular machines, or decentralized production cells.

Key contributions

- Human–Machine Collaboration: Smart wearables and force sensors ensure safe interaction between humans and robots in shared workspaces.
- Real-Time Coordination: Sensors communicate machine states, task completions, and environmental changes, allowing machines to synchronize dynamically.
- Interoperability: Standardized interfaces and protocols allow equipment from different vendors to work together, accelerating integration.

This collaboration leads to flexible, reconfigurable production lines that can adapt to changing product demands or customizations.

5.3.3.2 Data-Driven Decision Support

By providing high-resolution, contextualized data, smart sensors form the basis for intelligent decisions across manufacturing operations.

Examples

- Production Optimization: Real-time data helps optimize machine parameters for energy efficiency, material usage, and cycle times.
- Quality Assurance: Continuous monitoring ensures early detection of deviations or faults, allowing quick interventions.
- Supply Chain Resilience: Data on inventory levels, material flows, and machine availability supports predictive logistics and procurement planning.

This intelligence is often provided through dashboards, alerts, and automated reports, transforming how operational and strategic decisions are made.

5.3.3.3 Worker Safety and Environmental Monitoring

Smart sensors play an important role in ensuring worker safety and environmental compliance.

Worker safety applications

- Wearable Sensors: Track biometric parameters such as heart rate, fatigue, or exposure to harmful substances.
- Proximity Sensors: Detect worker location and halt machinery if someone enters a danger zone.
- Fall and Posture Detection: Alert supervisors if unsafe behavior or accidents occur.

Environmental monitoring applications

- Air Quality Sensors: Monitor Volatile Organic Compounds (VOCs), particulate matter, and toxic gases.
- Noise and Vibration Sensors: Ensure compliance with occupational health regulations.
- Thermal Sensors: Prevent overheating in critical infrastructure.

  These systems contribute to a safer, more compliant, and more productive work environment, reducing risks and enhancing workforce wellbeing.

5.4 Highlights

This chapter explored the significant role of sensors in IM, highlighting the developments from conventional to smart sensors and their expanding capabilities in the context of Industry 4.0. Key highlights of the chapter are summarized below:

Conventional sensors

a. Conventional sensors remain fundamental in capturing basic physical quantities like temperature, pressure, displacement, force, and proximity.
b. Their importance in process control, quality inspection, and inventory management remains vital in legacy and hybrid production environments.
c. Their simplicity, reliability, and cost-effectiveness make them suitable for structured, well-defined applications.

Soft sensors

a. Soft sensors, or virtual sensors, derive estimations or infer indirect measurements using mathematical models, AI algorithms, or data fusion techniques.
b. They offer advanced capabilities in predictive maintenance, real-time optimization, and inventory forecasting, often when physical sensors are impractical or expensive.
c. They enhance flexibility by integrating multiple data sources to generate a unified, accurate view of system performance.

Smart sensors

a. Smart sensors integrate sensing, processing, and communication into a single device.
b. They support edge computing, self-diagnosis, and IoT connectivity, facilitating real-time analytics and decentralized decision-making.
c. In IM, smart sensors are central to enabling collaborative manufacturing, data-driven operations, and worker/environmental safety.
d. Through interoperability with cloud platforms and CPSs, smart sensors are integral to digital twins and autonomous factories.

Exercises

P5.1 Define conventional sensors and list three common applications in manufacturing.

P5.2 What are soft sensors, and how do they differ from conventional sensors?

P5.3 Explain the concept of a smart sensor. What are its key components?

P5.4 List three advantages of smart sensors in an Industry 4.0 environment.

P5.5 What role do soft sensors play in predictive maintenance?

P5.6 Discuss the benefits and limitations of using conventional sensors in a modern manufacturing setup.

P5.7 How do smart sensors contribute to the creation and functioning of digital twins in intelligent factories?

P5.8 Evaluate the potential challenges of integrating soft sensors into existing industrial control systems.

P5.9 Describe how IoT connectivity enhances the effectiveness of smart sensors in real-time manufacturing processes.

P5.10 Compare and contrast the data output and processing requirements of conventional, soft, and smart sensors.

Notes

1. By Harke - Own work, Public Domain, https://commons.wikimedia.org/w/index.php?curid=6567327
2. By Marián Hubinský - Own work, CC BY-SA 4.0, https://commons.wikimedia.org/w/index.php?curid=135832024
3. By Ansgar Hellwig - photo taken with Canon PowerShot G3, CC BY-SA 2.0 de, https://commons.wikimedia.org/w/index.php?curid=242719
4. By Pleriche - Own work, CC BY-SA 4.0, https://commons.wikimedia.org/w/index.php?curid=61871859
5. What Is A Sensor, Different Types Of Sensors
6. By Jack LaRosa - Photographed and uploaded by me., Public Domain, https://commons.wikimedia.org/w/index.php?curid=4479143
7. By © Nevit Dilmen, CC BY-SA 3.0, https://commons.wikimedia.org/w/index.php?curid=30560805
8. By Ulfbastel, CC BY-SA 3.0, https://commons.wikimedia.org/w/index.php?curid=9061829

Part 2

Intelligent Manufacturing

Transcending the Notions

6

Role of Decision Intelligence and Machine Learning in Intelligent Manufacturing

6.1 Introduction to Decision Intelligence in Manufacturing

6.1.1 Understanding Decision Intelligence

Decision Intelligence (DI) represents a transformational approach in modern decision-making, combining the power of data science, AI, and behavioral science. It includes a framework that allows organizations to model decisions explicitly, understand the inputs and outcomes, and simulate potential consequences. Unlike traditional analytics, DI aims to guide what should happen next, based on contextual insights and predictive models.

DI is significant in complex environments like manufacturing, where decisions span operational efficiency, supply chain dynamics, resource allocation, and product quality. By creating structured decision-making processes that can adapt and learn from new data, DI enables proactive and informed decision-making across various organizational layers.

6.1.2 Application toward Manufacturing

Manufacturing processes today generate enormous volumes of data, from IoT sensors on the shop floor to ERP systems managing logistics. DI frameworks leverage this data to enhance agility and responsiveness. For instance, when a critical machine shows early signs of wear, DI systems can evaluate repair or replacement decisions considering cost, availability of parts, production schedules, and downstream impacts. Applications of DI in manufacturing include:

- Production Planning: Adaptive scheduling algorithms based on current demand, inventory levels, and machine availability.
- Supply Chain Optimization: Dynamic rerouting and procurement strategies in response to disruptions.
- Quality Management: Decision models to quickly respond to defects and process variations.
- Energy Management: Optimizing usage patterns based on forecasts and cost-benefit analyses.

DOI: 10.1201/9781003485186-8

6.1.3 Challenges and Benefits

The challenges of DI include:

- Data Quality and Integration: Manufacturing data often consists of data from systems with varied formats.
- Cultural Resistance: Transitioning from intuition-based to data-driven decisions can encounter resistance from experienced operators.
- Complexity of Modeling: Accurate decision models require a deep understanding of interdependencies and constraints.

The benefits of DI include:

- Faster decision cycles wherein real-time analytics enable timely interventions.
- Improved resource utilization by optimized workflows and inventory management, thereby reducing waste.
- Competitive advantage in terms of enhanced agility, quality, and responsiveness that distinguish leaders from laggards in the market.

6.2 Key Concepts in Decision Intelligence

6.2.1 Decision Support Systems (DSS)

These are interactive software-based systems designed to support complex decision-making. DSS in manufacturing integrates databases, models, and user interfaces to ease operational and strategic decisions. Modern DSS includes capabilities like real-time data visualization, scenario analysis, and prescriptive analytics.

6.2.2 Data Collection and Analysis

A critical foundation of DI is reliable and timely data. Data is collected from multiple sources such as:

- Operational systems such as ERP, MES, and SCADA
- Sensors and IoT devices like real-time environmental and machine data
- Human inputs such as maintenance logs and shift reports.

Analysis techniques for DI include descriptive statistics, root-cause analysis, and control charting, often integrated with AI for pattern recognition.

6.2.3 Predictive Analysis and Forecasting

Predictive analytics uses historical data and statistical algorithms to forecast future outcomes. Techniques such as regression, time-series analysis, and neural networks enable manufacturers to predict demand, anticipate equipment failures, and estimate lead times and process variations.

6.2.4 Quality Control and Defect Detection

DI frameworks help identify the root causes of quality issues by mapping decision trees and causal networks. Real-time image processing, anomaly detection, and feedback loops improve accuracy and reduce response time.

6.2.5 Challenges and Considerations

The challenges of DI include several key aspects. One major concern is model interpretability, as complex AI models often function as "black boxes," making it difficult to understand how decisions are derived. Another challenge is real-time processing, which involves ensuring low latency in decision-making, especially when dealing with high volumes of data. Lastly, scalability poses a significant hurdle, as extending DI models across multiple facilities or product lines requires robust infrastructure and consistent performance across varied environments.

6.2.6 Future Trends and Perspectives

Future trends and perspectives in DI point toward increasingly advanced and responsible applications. One emerging trend is the development of "cognitive decision systems," which integrate DI with natural language understanding to enable more intuitive and context-aware interactions. Another significant direction is the rise of "autonomous factories," where DI-enabled systems can make independent operational decisions with minimal human intervention, enhancing efficiency and responsiveness. Additionally, there is a growing focus on "ethical DI models," which prioritize transparency and fairness, aiming to eliminate bias and ensure that decision-making mechanisms are accountable and aligned with societal values.

6.3 Introduction to Machine Learning

ML represents one of the most transformative technological developments of the 21st century. As a subfield of AI, ML equips computers with the ability to learn patterns and make decisions based on data, without being explicitly programmed for each specific task. In the context of IM, ML serves as the backbone of predictive systems, quality control mechanisms, process optimization, and autonomous systems.

Unlike traditional rule-based programming, where specific instructions are hardcoded, ML systems generalize from examples. When provided with historical manufacturing data such as machine temperature logs, vibration levels, defect reports, or production throughput, ML models can learn relationships between input variables and desired outcomes. This learned understanding enables the systems to make predictions or classifications on new, unseen data.

6.3.1 Understanding Basic Concepts

To build and deploy ML solutions in manufacturing environments, a structured workflow commonly known as the ML pipeline is essential. This pipeline consists of several crucial stages:

- Data Acquisition: This step involves gathering raw data from various sources such as sensors, MES, ERP platforms, and operator logs. Data may include time-series data from machines, environmental readings, image data for visual inspection, or structured databases of production records.
- Preprocessing: Raw data is often noisy, incomplete, and inconsistent. Preprocessing steps may involve data cleaning (e.g., removing outliers, handling missing values), normalization (scaling numeric values to standard ranges), and transformation (e.g., converting categorical data to numerical format). This ensures that data fed into ML models is reliable and standardized.
- Feature Engineering: This step involves transforming raw data into meaningful input variables (features) that improve model performance. For example, deriving "rate of temperature change" or "average vibration over 10 minutes" can offer better insights than using raw sensor readings alone.
- Model Training and Evaluation: In this phase, various algorithms are trained on historical data to learn patterns. The dataset is typically divided into training, validation, and test sets. Evaluation metrics such as accuracy, precision, recall, F1-score, or mean squared error (depending on the task) are used to measure model performance and avoid overfitting.
- Deployment and Monitoring: Once a model is validated, it is deployed into the production environment. Continuous monitoring is critical to ensure that the model remains accurate over time. In manufacturing, models can degrade due to data drift or changes in operating conditions, necessitating periodic retraining.

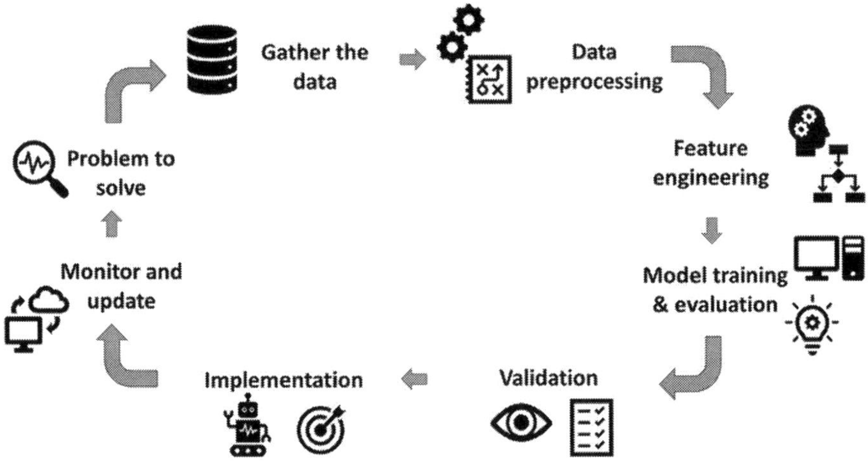

FIGURE 6.1
A simplified sketch of steps in ML.

A simplified sketch of the steps in ML[1] is shown in Figure 6.1.

6.3.2 Application toward Manufacturing

ML finds application across a wide spectrum of manufacturing use cases, such as

- Anomaly Detection: ML algorithms analyze sensor streams in real time to detect deviations from normal operating behavior. For instance, a model trained on vibration and temperature data can alert operators if a machine is likely to fail, long before a human might notice the signs. Unsupervised learning or autoencoder neural networks are typically used in such scenarios.
- Demand Forecasting: Historical sales data, combined with external factors like seasonality and market trends, can be fed into regression models or ensemble methods (e.g., random forests and gradient boosting) to forecast future product demand. Accurate forecasts improve inventory management and reduce waste.
- Process Optimization: Reinforcement learning algorithms can learn to adjust manufacturing parameters such as speed, feed rate, and temperature by maximizing a reward function like output quality or energy efficiency. This is particularly useful in continuous processes like chemical production or semiconductor manufacturing.
- Robotics and Automation: In modern factories, robots equipped with ML models are used for tasks such as visual inspection, object manipulation, and autonomous navigation. Convolutional Neural

Networks (CNNs) help robots identify objects or defects, while recurrent or attention-based networks assist in sequencing actions. Reinforcement learning enables adaptive behavior in complex environments.

6.3.3 Types of Machine Learning Algorithms

ML algorithms are categorized based on the nature of the data they are exposed to during training and the kind of tasks they are designed to perform. Understanding these categories helps manufacturing professionals choose the right approach for different industrial problems.

6.3.3.1 Supervised Learning

Supervised learning is the most commonly used category of ML. In this approach, the model is trained on a dataset that includes both inputs and known outputs (labels). The algorithm learns to map the inputs to the correct outputs and generalizes this learning to make predictions on new, unseen data.

In Manufacturing, supervised learning is extremely useful when historical labeled data is available, e.g., sensor data tagged with "normal" or "faulty" labels, or records of machines annotated with their lifespan.

The applications of supervised learning are as follows:

a. Classification Tasks:
   - Detecting defective products using image data or sensor readings.
   - Classifying fault types based on vibration signals.
   - Sorting raw materials or components into categories using visual or physical properties.
   - Example Algorithms: Decision trees, Random forests, k-nearest neighbors (k-NN), SVM, and CNNs.

b. Regression Tasks:
   - Predicting numeric values such as machine wear levels, product thickness, temperature drift, or time-to-failure.
   - Estimating the optimal process parameters for desired product quality.
   - Example Algorithms: Linear regression, Ridge regression, Gradient boosting machines (GBMs), Deep neural networks.

One of the key advantages of supervised learning is that, in some cases, such as with decision trees, it offers easier model interpretability, allowing engineers to understand and trust the decision-making process. However, supervised learning also comes with limitations. It requires a large volume of labeled data, which can be both expensive and time-consuming to gather in complex manufacturing environments. This dependency on labeled datasets often becomes a bottleneck, especially when scaling solutions across varied production scenarios.

6.3.3.2 Unsupervised Learning

In unsupervised learning, the model receives input data without any associated labels. The algorithm explores the data and discovers hidden patterns, structures, or relationships on its own.

In Manufacturing, unsupervised learning is useful when labeled data is unavailable or impractical to obtain, such as in early-stage diagnostics, process exploration, or quality clustering.

The applications of unsupervised learning are as follows:

a. Clustering:
   - Grouping similar products, components, or production batches based on features like dimensional measurements, color consistency, or texture.
   - Segmenting customers or suppliers in supply chain analytics.
   - Example Algorithms: k-Means, DBSCAN, Hierarchical clustering.

b. Dimensionality Reduction:
   - Simplifying complex datasets with hundreds or thousands of variables (e.g., hyperspectral images or multichannel sensor readings).
   - Helps improve visualization and reduce computational complexity.
   - Techniques include principal component analysis (PCA).

c. Anomaly Detection:
   - Identifying deviations from normal patterns without needing labeled faults.
   - Particularly useful for rare failure events or process outliers.
   - Algorithms: Isolation forests, One-Class SVM, Autoencoders.

The key advantages of unsupervised learning include the ability to uncover previously unknown insights and support exploratory data analysis, making it ideal for identifying anomalies, clusters, or trends within complex datasets. However, unsupervised learning also presents challenges. Evaluating model performance is inherently difficult due to the absence of labeled outcomes, and the results it produces often require significant domain expertise to interpret accurately. This reliance on expert judgment can limit its direct applicability in automated decision-making scenarios.

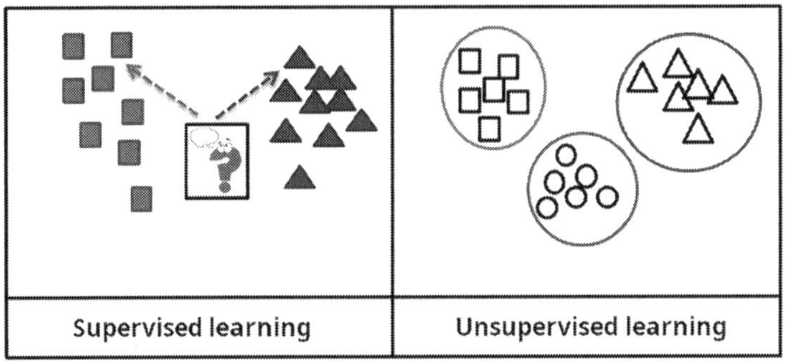

FIGURE 6.2
Difference between supervised and unsupervised learning.

A simple sketch[2] showing the difference between supervised and unsupervised learning is shown in Figure 6.2.

6.3.3.3 Reinforcement Learning

Reinforcement learning (RL) is a unique approach where an agent learns to make decisions by interacting with its environment. The agent takes actions, observes the outcomes, and receives rewards or penalties based on performance. Over time, it optimizes its behavior to maximize cumulative reward (see the example shown in Figure 6.3[3]).

In Manufacturing, RL is best suited for dynamic and complex environments where real-time decisions affect future states such as in autonomous systems, adaptive scheduling, or closed-loop control.

The applications of RL include:

a. Robotic Path Planning:
    - Teaching robots to navigate through factory floors, avoid obstacles, and adapt to new layouts using trial-and-error learning.
    - Algorithms include Q-Learning, Deep Q-Networks (DQN), and Proximal Policy Optimization (PPO).

b. Adaptive Control Systems:
    - Adjusting process parameters such as pressure, temperature, or flow rate in real time based on changing conditions.
    - Examples include chemical reactors, CNC machine tuning, or injection molding optimization.

c. Smart Scheduling:
    - Learning to sequence jobs or allocate machines optimally to minimize makespan, energy consumption, or downtime.

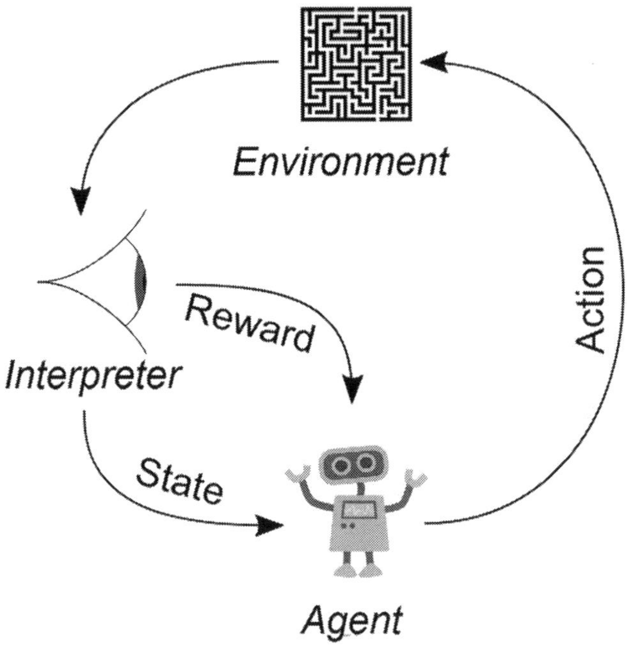

FIGURE 6.3
Reinforcement learning principle.

- RL agents consider multiple objectives and long-term outcomes, not just immediate efficiency.

One of the key advantages of RL is its ability to discover novel solutions that may not be evident to human planners, making it a valuable tool for process optimization and autonomous control. However, reinforcement learning also has notable limitations. It typically requires extensive computational resources and considerable exploration time to converge on effective strategies. Moreover, without the use of controlled simulations, there is a risk of the system making unsafe or suboptimal decisions during the training phase, which can be problematic in real-world industrial settings.

6.4 Key Concepts in Machine Learning for Manufacturing

This section explores the foundational elements and advanced applications of ML in manufacturing environments, highlighting how each contributes to creating resilient, efficient, and high-performing industrial systems.

6.4.1 Data Collection and Preprocessing

Data is the raw material of ML, and its quality directly impacts the reliability, accuracy, and applicability of the resulting models. In manufacturing, data originates from different sources such as IoT sensors, SCADA systems, ERP software, vision systems, and human-entered logs. Before this data can be used for training ML models, it must be preprocessed to ensure consistency, relevance, and readiness.

Key preprocessing steps include:

- Handling Missing Data: Real-world sensor streams and operational records often contain gaps due to transmission errors, machine downtime, or manual oversights. Techniques such as mean imputation, forward filling, or model-based inference (e.g., k-NN or regression) are used to fill in missing values.
- Data Normalization and Scaling: Different features often operate on different scales (e.g., temperature in Celsius vs. pressure in Pascals). Standardization (z-scores) or normalization (min-max scaling) ensures that no feature dominates others purely due to scale.
- Dimensionality Reduction: High-dimensional data, such as multichannel sensor arrays or high-resolution images, can introduce noise and overfitting. Techniques like PCA and t-distributed Stochastic Neighbor Embedding (t-SNE) reduce dimensionality while preserving the most informative patterns.
- Feature Selection and Engineering: Selecting the most relevant features, or engineering new ones (e.g., rolling averages, deltas, ratios), enhances model interpretability and performance. Feature importance techniques like SHAP, permutation importance, or domain expert input are often used here.

This preprocessing pipeline lays the groundwork for robust and interpretable ML systems in the factory setting.

6.4.2 Predictive Maintenance

Predictive Maintenance (PdM) is among the most high-impact applications of ML in manufacturing. It shifts maintenance from reactive (after failure) and preventive (scheduled) approaches to data-driven, condition-based interventions.

Key techniques of PdM include:

- Time-Series Analysis: Sensor data like vibration, temperature, and current are analyzed to detect degradation trends. ML models learn temporal patterns and anomalies that precede equipment failure.

- Classification Models: Supervised learning models, such as decision trees, random forests, and neural networks, can classify machines as "healthy" or "at risk" based on labeled historical data.
- Survival Analysis: Statistical methods (e.g., Cox Proportional Hazards and Kaplan–Meier estimators) estimate the RUL of a component or machine, helping plan timely interventions.

The benefits of PdM include reduction in unplanned downtime, prolonged asset lifespan, lower maintenance costs, and better spare parts inventory planning. By predicting failures before they happen, ML-driven PdM improves plant reliability and operational continuity.

6.4.3 Quality Control and Defect Detection

In high-throughput manufacturing environments, traditional quality checks are often too slow or inconsistent. ML-based systems bring real-time, scalable, and non-invasive quality assurance mechanisms. Key techniques for quality control and defect detection include:

- Computer Vision with CNNs: CNNs are highly effective at image-based inspection. They can detect surface defects, misalignments, scratches, or foreign particles with pixel-level precision.
- Unsupervised Anomaly Detection: For processes lacking labeled defect data, unsupervised methods such as k-means clustering, Isolation Forests, or Autoencoders can learn "normal" operating states and flag deviations.
- Multivariate Analysis: Manufacturing often involves multiple interdependent variables. ML models can detect when a combination of otherwise normal variables leads to a defect, a scenario difficult to detect using traditional Statistical Process Control (SPC) methods.

The applications for quality control include:

- Visual inspection in electronics, automotive, and packaging
- Real-time defect segregation on conveyor belts, and
- In-line monitoring for continuous production processes.

6.4.4 Quality Optimization

Beyond defect detection, ML also plays a proactive role in improving process quality by discovering optimal settings and process configurations. The optimization techniques used in quality are as follows:

- Bayesian Optimization: Ideal for tuning parameters (e.g., temperature, speed, viscosity) in expensive or slow experiments, it balances exploration and exploitation.
- GA: Evolution-inspired algorithms that search for the best combination of parameters by simulating natural selection.
- RL: Used for dynamic environments, RL agents learn to adjust process parameters by receiving reward signals based on product quality or throughput.

Example use cases for quality optimization include:

- Tuning injection molding parameters to reduce warpage
- Optimizing chemical reaction conditions for yield, and
- Determining optimal CNC cutting speeds for quality and tool life.

ML-driven optimization not only boosts quality but also supports sustainability by reducing waste and energy consumption.

6.4.5 Human–Machine Collaboration

Rather than replacing humans, ML augments their capabilities by providing decision support, contextual awareness, and automation assistance. The integration avenues are many with few mentioned below:

- Intelligent Dashboards: ML-powered dashboards prioritize alerts, surface anomalies, and recommend actions, helping managers and operators make quicker and better-informed decisions.
- AR: Combined with ML, AR overlays real-time insights on physical equipment, guiding technicians during maintenance or troubleshooting.
- Collaborative Robotics (Cobots): ML enables robots to learn from human demonstrations, adapt to operator behavior, and safely share workspaces.

This human–machine synergy increases productivity, enhances safety, and reduces the cognitive burden on human operators.

6.4.6 Challenges and Considerations

Despite the benefits, several challenges can hinder the effective deployment of ML in manufacturing such as

- Data Drift and Model Degradation: As equipment ages or production conditions change, models trained on past data may become obsolete. Continuous monitoring and retraining of pipelines are necessary.
- Domain-Specific Feature Engineering: ML models often require features that encapsulate domain knowledge (e.g., torque per unit load). Collaboration with process engineers is essential.
- Generalization across Products or Lines: Models trained on one product line may not perform well on another due to subtle variations. Transfer learning and federated learning are promising solutions.
- Infrastructure Limitations: Real-time ML demands low-latency, high-bandwidth data infrastructure that may not exist in legacy systems.

These considerations must be addressed through robust design, testing, and operational integration strategies.

6.4.7 Future Trends and Perspectives

The landscape of ML in manufacturing is evolving rapidly, driven by new technologies and industrial priorities. The future trends and perspectives include:

- Edge Computing: Moving ML inference to edge devices (e.g., PLCs and smart sensors) enables real-time, low-latency decision-making without reliance on cloud connectivity.
- Explainable AI (XAI): Increasing demand for transparency and trust in ML models is pushing adoption of interpretable techniques that justify predictions and decisions.
- Blockchain Integration: Combining ML with blockchain ensures data provenance, security, and traceability, which are critical in regulated industries like food, pharmaceuticals, and aerospace.
- Self-Learning Factories: The vision is of systems that continuously improve without human intervention, where ML models are retrained in real time, and decisions adapt autonomously to new situations.

As these technologies mature, manufacturers will be empowered to create truly intelligent systems that are not only reactive but predictive, adaptive, and self-optimizing.

6.5 Integration of Machine Learning and Decision Intelligence

The integration of ML and DI represents a powerful convergence of prediction and action. While ML offers the ability to uncover patterns, forecast future outcomes, and extract insights from vast datasets, it often stops short of determining what should be done with those insights. This is where DI steps in, i.e., bridging the gap between what is likely to happen and what should be done in response.

6.5.1 Comparative Analysis of Machine Learning and Decision Intelligence

ML and DI serve distinct but complementary roles within IM systems. Understanding their fundamental differences is crucial for designing systems that go beyond prediction and drive intelligent action (Table 6.1).

Illustrative Example

In a manufacturing context, ML might generate an output indicating that a specific CNC machine is likely to fail within the next 48 hours. Building on this, a DI system takes the prediction further by evaluating it against various operational factors such as upcoming production deadlines, availability of replacement parts, technician schedules, and the potential cost implications of downtime. Based on this holistic assessment, the DI system recommends the most optimal course of action: whether to allow the machine to complete its current tasks, shift the workload to other production lines, or initiate immediate preventive maintenance.

TABLE 6.1

Table Showing the Difference between ML and DI for Manufacturing

Criteria	ML	DI
Goal	Learn patterns from historical or real-time data	Make optimal decisions based on data, context, and desired outcomes
Scope	Narrow, task-specific (e.g., defect detection, demand forecasting)	Broad, process-oriented (e.g., resource allocation, production planning)
Input	Purely data-driven	Data combined with domain knowledge, objectives, constraints, and contextual information
Output	Prediction (e.g., failure probability, product quality estimate)	Actionable recommendation or decision (e.g., shut down machine and reassign workforce)

This distinction demonstrates that while ML is a necessary component for insight generation, it is DI that completes the decision loop by formulating and simulating alternative actions.

6.5.2 Exploring the Synergy between Machine Learning and Decision Intelligence

The full potential of IM is realized not by using ML or DI in isolation, but through their synergistic integration. This combination creates dynamic, context-aware systems capable of both anticipating problems and responding to them autonomously or semi-autonomously. The key synergistic advantages include:

a. Closed-Loop Decision Systems: Integrated ML-DI architectures enable closed feedback loops where the outcomes of decisions are monitored and used to continuously refine ML models. For instance, if a DI system schedules a maintenance task based on ML predictions, the actual effectiveness and timeliness of that intervention are fed back into the model to improve future predictions.
b. Context-Aware Decision Recommendations: ML models may predict "what" is likely to occur (e.g., an increase in product defect rate), but DI provides the "why" and "what now" by embedding operational constraints, priorities, and trade-offs. This transforms raw predictions into strategic choices prioritizing decisions that align with larger business goals like minimizing cost, maximizing throughput, or maintaining compliance.
c. Real-Time Operational Responsiveness: In a high-velocity manufacturing environment, delays in decision-making can lead to cascading failures or missed opportunities. Integrated systems can analyze streaming data, make predictions in real time, and immediately trigger appropriate actions such as diverting defective products, rerouting logistics, or dynamically adjusting machine parameters.

The following example shows the integrated workflow and highlights the operational intelligence that emerges from blending prediction with judgment.

- IoT sensors stream temperature and vibration data from a robotic assembly line.
- An ML model detects abnormal vibration patterns indicating potential bearing wear.

- DI module evaluates production schedules, remaining order volume, and technician availability.
- System decides to schedule downtime during the next planned shift change and orders replacement parts in advance to minimize disruption.

6.5.3 Benefits of Integration of Machine Learning and Decision Intelligence

The fusion of ML and DI in IM delivers transformative benefits that enhance every layer of operations, from the shop floor to executive planning. The benefits are many with few listed as follows:

a. Reduced Downtime through Intelligent Intervention: ML enables early detection of risks, while DI ensures those risks are mitigated in a cost-effective and timely manner. This proactive maintenance and resource allocation strategy significantly lowers unplanned downtime, leading to more predictable operations.
b. Seamless Coordination between Planning and Execution: Integrated ML-DI systems create adaptive workflows where plans are continuously revised based on real-time data and contextual understanding. This creates a living digital twin of the manufacturing process, always current, always responsive.
c. Higher Quality and Reduced Waste: ML algorithms trained on historical quality data can anticipate conditions that lead to defects. DI frameworks use this information to optimize process parameters or halt production when tolerances are at risk. This minimizes scrap, rework, and warranty costs.

6.6 Highlights

- DI is transforming manufacturing by structuring, simulating, and optimizing complex decisions.
- ML offers the predictive muscle to foresee patterns, failures, and opportunities.
- Their integration creates adaptive, resilient, and intelligent systems capable of autonomous operation.
- Adoption challenges remain but are being overcome through innovation and upskilling.

Exercises

P6.1 What is Decision Intelligence, and how does it differ from traditional analytics?

P6.2 List any three common challenges in implementing Decision Intelligence in manufacturing.

P6.3 What are the main steps in a typical Machine Learning pipeline?

P6.4 Explain the difference between supervised and unsupervised learning with one example each from manufacturing.

P6.5 What is predictive maintenance, and why is it important in smart factories?

P6.6 Mention two future trends in Machine Learning for manufacturing and briefly describe them.

P6.7 Discuss how Machine Learning and Decision Intelligence work together to enhance manufacturing operations. Provide relevant examples.

P6.8 Explain the role of Data Collection and Preprocessing in Machine Learning for manufacturing. Why is it critical for model success?

P6.9 Describe in detail the various applications of Machine Learning in the manufacturing industry, such as anomaly detection, robotics, and process optimization.

P6.10 What is Quality Optimization in ML-based manufacturing? Explain any two techniques used for it.

P6.11 Provide a detailed comparison between Machine Learning and Decision Intelligence. Include differences in goals, inputs, outputs, and scope.

P6.12 How does Human–Machine Collaboration work in IM systems? Describe how ML and AR tools contribute to better decisions on the shop floor.

Notes

1 Ledziński, Ł., & Grześk, G. (2023, April). Artificial intelligence as an emerging tool for cardiologists. *Medical Sciences Forum*, 21(1), 15.

2 By Balkiss.hamad - File:Apprentissage_Supervisé_Vs_Non_Supervisé.png(Derived version translated to English by Alenoach, CC BY-SA 4.0, https://commons.wikimedia.org/w/index.php?curid=155627022

3 By Megajuice - Own work, CC0, https://commons.wikimedia.org/w/index.php?curid=57895741

7

Immersive Technologies in Intelligent Manufacturing

7.1 Introduction

As manufacturing undergoes a significant transformation in the era of Industry 4.0, immersive technologies are becoming critical enablers of IM. These technologies offer a significant shift in how humans interact with machines, data, and environments by merging the physical and digital realms. The deployment of VR, AR, and Mixed Reality (MR) is facilitating smarter, faster, and safer manufacturing operations. They empower decision-makers, optimize workflows, and address challenges such as skill gaps, operational inefficiencies, and the complexity of modern production systems.

The manufacturing sector, characterized by its drive for precision, efficiency, and adaptability, is an ideal environment for immersive technology adoption. Through interactive simulations, real-time overlays, and immersive training platforms, manufacturers are modifying the factory floor as a hybrid of physical and digital worlds. This convergence encourages enhanced human-machine collaboration and lays the foundation for the FoF.

7.2 Types of Immersive Technologies

Immersive technologies are typically categorized into three major types: VR, AR, and MR. Each offers unique capabilities and levels of interaction between the real and virtual worlds.

7.2.1 Virtual Reality (VR)

VR is a fully immersive technology that transports users into a computer-generated three-dimensional environment, effectively isolating them from the physical world. Using specialized hardware such as VR headsets (as shown in Figure 7.1), motion controllers, and haptic devices (as shown in Figure 7.2), users can explore, interact with, and manipulate digital environments in real time. This immersive experience enhances spatial awareness,

DOI: 10.1201/9781003485186-9

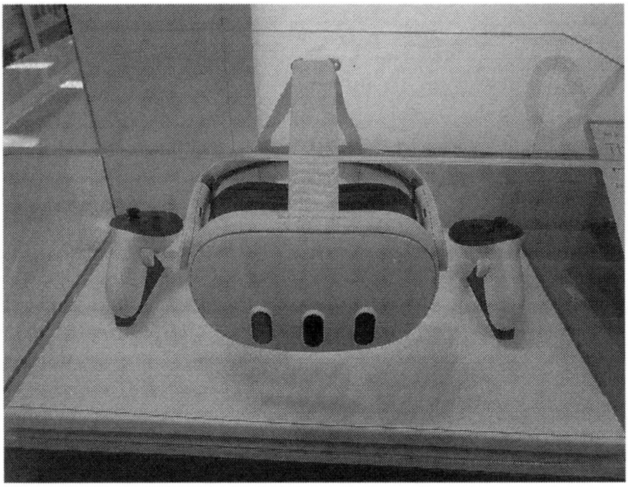

FIGURE 7.1
A VR headset from Metaquest. (By IHazACatNamedMax - Own work, CC BY-SA 4.0, https://commons.wikimedia.org/w/index.php?curid=139735935.)

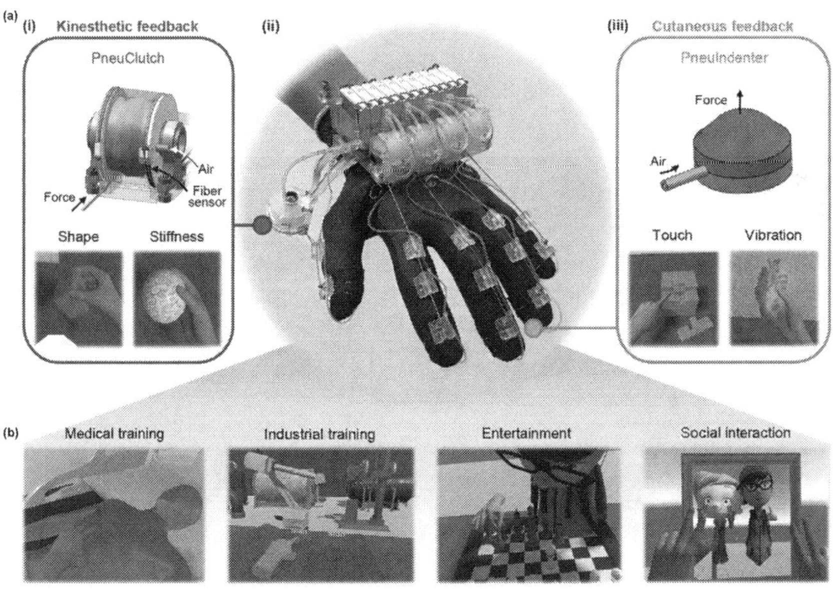

FIGURE 7.2
Haptic device for interacting with digital environment. (a) (i) The kinesthetic feedback module delivers the shape and stiffness information of virtual objects. (ii) Image of the device worn on hand. (iii) The feedback module provides touch and vibration feedback. (b) Use of devices enhancing virtual medical training, industrial training, entertainment, and socialization. (Qi, J., Gao, F., Sun, G., Yeo, J. C., & Lim, C. T. (2023). HaptGlove – untethered pneumatic glove for multimode haptic feedback in reality–virtuality continuum. *Advanced Science, 10*(25), 2301044.)

decision-making, and learning by simulating real-world scenarios with high fidelity and precision.

In the context of IM, VR is changing how engineers, designers, operators, and managers engage with complex systems and processes. Unlike traditional 2D models or static simulations, VR offers an embodied experience, allowing stakeholders to "walk through" virtual factories, evaluate product ergonomics, or simulate assembly lines, all without incurring physical or operational costs.

Core characteristics of VR include:

- Full Environment Simulation: VR creates a completely virtual space detached from the real world.
- Interactive Feedback Loops: Real-time interaction allows users to manipulate virtual objects and receive immediate feedback.
- Immersive Learning: Users gain experiential understanding through trial and error in a risk-free setting.

7.2.2 Augmented Reality (AR)

AR is a transformative immersive technology that overlays digital content such as 3D models, contextual data, instructions, or alerts onto the user's physical environment in real time. Unlike VR, which immerses the user in a fully virtual space, AR enhances the real world by adding layers of actionable information without disconnecting users from their surroundings. This dual presence of digital and physical elements enables more informed, intuitive, and efficient decision-making on the shop floor.

AR systems typically utilize devices such as smartphones, tablets, AR glasses [e.g., Microsoft HoloLens (as shown in Figure 7.3) or Magic Leap (as shown in Figure 7.4)], or head-mounted displays. These devices use cameras, sensors, and computer vision algorithms to detect the user's environment, recognize objects, and precisely position digital overlays in context.

In the context of IM, AR acts as a real-time interface between humans, machines, and data. It supports smarter workflows, improves operational accuracy, and enhances the cognitive capabilities of workers by delivering information when and where it's needed.

Core characteristics of AR include:

- Context-Aware Visualization: AR systems use spatial mapping and object recognition to present digital data that is directly tied to the physical world.
- Hands-Free Operation: Wearable AR devices allow workers to access critical information while keeping their hands free for tasks.
- Real-Time Interaction: Information is updated dynamically based on user movement, machine status, or system inputs.

FIGURE 7.3
Microsoft HoloLens. A head-mounted AR device. (By Kai Kowalewski - Own work, CC BY-SA 4.0, https://commons.wikimedia.org/w/index.php?curid=48112096.)

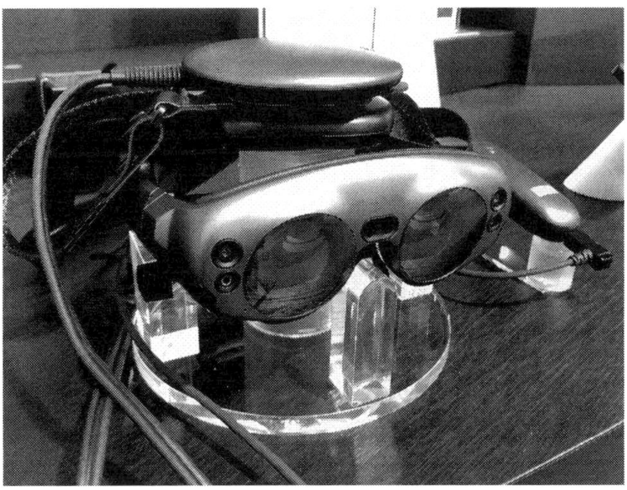

FIGURE 7.4
Magic Leap. (By KKPCW - Own work, CC BY-SA 4.0, https://commons.wikimedia.org/w/index.php?curid=83055632)

7.2.3 Mixed Reality (MR)

MR represents the most sophisticated and dynamic point of immersive technology methodologies, seamlessly blending the physical and digital worlds into a single, interactive environment. Unlike VR and AR, MR enables real-time interaction between physical and virtual elements. Users can manipulate digital objects as if they were real, and those digital objects

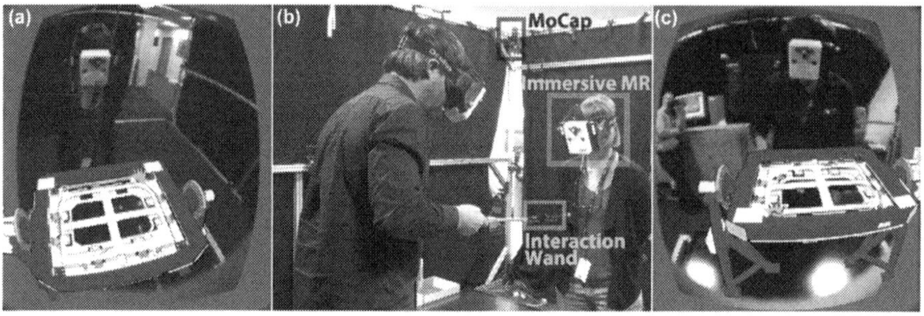

FIGURE 7.5
MR setup. (a) Trainer's view, (b) experimental environment with sensors and operators, and (c) participant's view. (Gonzalez-Franco, M., Pizarro, R., Cermeron, J., Li, K., Thorn, J., Hutabarat, W., ... & Bermell-Garcia, P. (2017). Immersive mixed reality for manufacturing training. *Frontiers in Robotics and AI*, *4*, 3.)

can respond to the environment, adapting their behavior based on spatial context, user input, or machine data. An example of MR setup is shown in Figure 7.5.

MR is powered by advanced spatial computing, combining ML, computer vision, sensor fusion, and 3D mapping to recognize, understand, and respond to real-world geometry and interactions. In the context of IM, MR serves as a pivotal interface between digital ecosystems (e.g., digital twins, IoT networks, and AI analytics) and the human workforce.

Core characteristics of MR include:

- Bidirectional Interaction: Virtual elements respond to changes in the physical environment (e.g., lighting, movement, and gestures).
- Persistence and Anchoring: Digital content can be spatially anchored to specific real-world objects or environments.
- Context-Aware Adaptability: MR applications adapt in real time based on contextual input from sensors, user actions, or machine data.

7.3 Applications in Intelligent Manufacturing

From product development to shop-floor execution, immersive technologies are enhancing collaboration, accelerating innovation, and optimizing performance. This section explores four key areas where these technologies are delivering measurable value in SM environments.

7.3.1 Product Design and Prototyping

The early phases of product development, namely, ideation, concept validation, and prototyping, are critical to determining time-to-market, cost, and product success. Immersive technologies, particularly VR and MR, enable a radical shift from traditional approaches to collaborative, real-time innovation.

Applications for product design and prototyping include:

- Virtual Prototyping: Designers can create fully interactive 3D models (as shown in Figure 7.6) in VR/MR environments, allowing cross-functional teams to test form, fit, and function before any physical prototype is produced.
- Human-Centered Design: MR allows designers and ergonomists to visualize how products will be used in real environments, supporting usability, accessibility, and safety considerations.
- Design Collaboration: Teams from multiple geographies can meet in shared virtual spaces to iterate on designs, annotate models, and review engineering changes – eliminating delays and versioning issues.

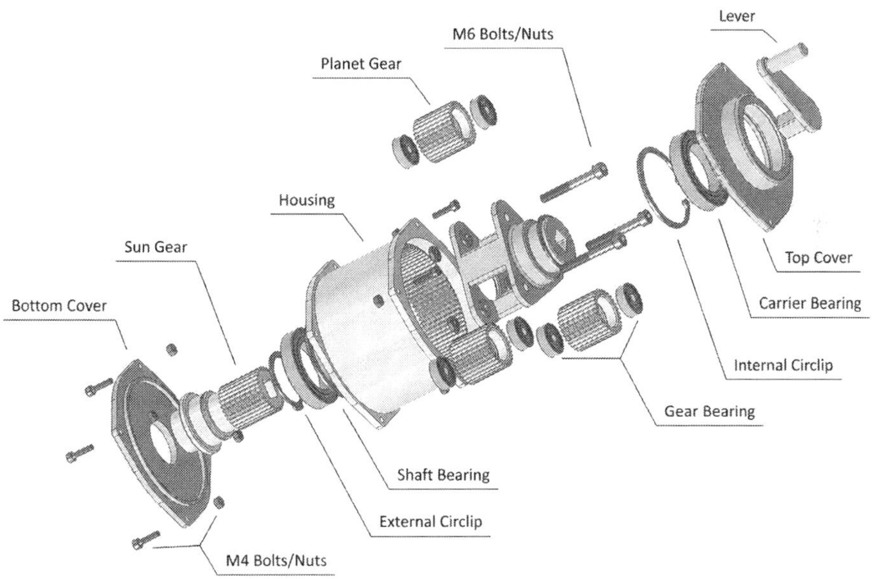

FIGURE 7.6
3D model of a gear assembly used for design validation using MRtools. (Kolla, S., Kumar, A. A., & Plapper, P. (2023). 8 mixed reality for Industry 4.0 in manufacturing systems. In Uzair Khaleeq uz Zaman, Ali Siadat, Aamer Ahmed Baqai, Kanwal Naveed, and Atal Anil Kumar (Eds.), *Handbook of Manufacturing Systems and Design: An Industry 4.0 Perspective*, 123. Boca Raton, FL: Taylor & Francis (CRC Press imprint).)

7.3.2 Training and Skill Development

One of the most powerful uses of immersive technologies lies in training and workforce upskilling. Traditional training methods often lack interactivity, are costly to scale, and may fail to simulate real-world complexity. VR, AR, and MR offer experiential learning environments that improve knowledge retention, safety, and task proficiency.

Applications for training and skill development include:

- VR-Based Simulations: Workers can learn complex tasks such as CNC programming, machine operation, or hazardous procedures in safe, immersive environments that replicate real equipment and scenarios.
- AR-Enabled Job Aids: New operators receive contextual, step-by-step instructions through AR glasses, allowing them to perform tasks while learning on the job.
- MR-Based Collaborative Training: Teams can rehearse complex operations (e.g., aircraft maintenance and robotic assembly) in hybrid environments, where virtual components interact with physical tools.

7.3.3 Smart Factory Operations

The concept of smart factory is to develop a cyber-physical space where real-time data, adaptive systems, and autonomous processes converge. Immersive technologies enhance operational visibility and agility by providing context-aware, real-time interfaces to monitor and manage manufacturing systems.

Applications of smart factory operations include:

- AR Dashboards and Machine Interfaces: Operators view live machine performance, alerts, and key performance indices directly on equipment via AR overlays, enabling faster response to issues.
- MR-Driven Control Rooms: Supervisors can interact with 3D models of production lines, visualize IoT sensor data spatially, and simulate changes in throughput, layout, or scheduling.
- Remote Monitoring and Control: Technicians can inspect and control machines from remote locations using immersive environments linked to IoT and cloud platforms.

7.3.4 Quality Control and Inspection

In this era, quality assurance is about integrating real-time quality feedback throughout the production cycle. Immersive technologies play a critical role in visual verification, process compliance, and early error detection.

Applications for quality control and inspection include:

- AR-Assisted Inspections: Inspectors can compare physical assemblies against CAD models using AR overlays to spot deviations, missing components, or misalignments with micron-level accuracy.
- Defect Visualization: MR environments allow engineers to explore defects detected by vision systems, overlay root cause analysis, and simulate corrective actions.
- VR-Based Failure Simulations: Teams can recreate product failures or process anomalies in virtual settings, analyze outcomes, and adjust quality protocols accordingly.

7.4 Simulation and Digital Twins

Simulation and digital twin technologies are one of the key enablers of data-driven decision-making and real-time process optimization in IM. These technologies act like a bridge between the physical and digital worlds, helping manufacturers create virtual versions of their operations so they can test, monitor, and improve them with greater accuracy.

With technologies like IoT, AI, and cloud computing, simulations are now dynamic, interactive systems that update in real time to mirror what's actually happening on the factory floor. Digital twins, in particular, represent the evolution of simulation: they are living digital replicas of physical entities that receive continuous data input from the real world. As such, they not only simulate behavior but also adapt and predict outcomes based on real-time conditions. Together, simulations and digital twins empower organizations to shift from reactive to proactive manufacturing, enabling continuous improvement, cost efficiency, and strategic foresight.

7.4.1 Simulating Manufacturing Processes

Simulation in manufacturing refers to the creation of digital models that imitate the behavior of a manufacturing process, system, or entire facility. These models are used to evaluate performance, optimize operations, and test scenarios without interrupting real-world production. As immersive technologies become more integrated into simulation platforms, engineers and decision-makers can now interact with these virtual models through VR/MR environments, enhancing their ability to visualize, analyze, and optimize.

Types of simulations in manufacturing include:

- Discrete Event Simulation (DES): Models the operation of a system as a discrete sequence of events (e.g., material arrival, machine changeover).
- FEA: Simulates physical responses like stress, heat, and vibration in product and tool designs.
- CFD: Models airflow, fluid behavior, and thermal effects, often used in energy or HVAC component design.
- System Dynamics and Agent-Based Models: Evaluate complex system behaviors and human-machine interactions across large-scale production networks.

7.4.1.1 Immersive Integration

Immersive simulations offer visual fidelity and spatial interactivity. Engineers can virtually walk through factory layouts, test ergonomics, simulate bottlenecks, and conduct "what-if" scenario planning, all within MR environments. Operators can train on simulated equipment with realistic feedback and visual cues, reducing the learning curve and minimizing disruption to ongoing operations.

Key benefits of simulating manufacturing processes include:

- Design Validation: Simulations allow manufacturers to test process changes, new layouts, or automation sequences before implementation, reducing risk.
- Process Optimization: Engineers can identify bottlenecks, inefficiencies, and energy drains by running time-and-motion studies in simulated environments.
- Scenario Analysis: "What-if" modeling (e.g., equipment failure, demand surge) helps prepare for uncertainty and supports agile planning.
- Cost Savings: Eliminates the need for physical prototyping or trial runs, lowering investment risk and operational costs.

7.4.2 Predictive Maintenance

PdM is one of the most impactful applications of digital twins in IM, as highlighted in previous chapters. It involves using real-time data, historical trends, and ML algorithms to forecast equipment failures before they occur. By combining sensor-based monitoring with digital twin models, manufacturers can proactively plan maintenance schedules, reduce unplanned downtime, and extend asset lifecycles.

7.4.2.1 Digital Twin in Maintenance

A digital twin not only represents the physical structure of an asset but also its operational data including vibration, temperature, humidity, pressure, and usage patterns. Advanced PdM systems use this data to simulate future behavior, identify early warning signs of degradation, and even recommend corrective actions. Technicians equipped with AR headsets can:

- View equipment health metrics overlaid directly on physical assets.
- Receive alerts and recommended maintenance steps.
- Access guided repair instructions or connect with remote experts for assisted diagnostics.

MR environments play a transformative role in PdM by enabling real-time simulation of component failures. This allows maintenance teams to visualize the internal condition of machines and rehearse repair procedures before physically interacting with the equipment, improving both efficiency and safety.

The advantages of PdM are substantial. It leads to increased equipment uptime by proactively addressing potential failures before they result in unexpected breakdowns. Resource allocation is optimized, as maintenance is performed based on actual asset conditions rather than fixed schedules. PdM also enhances safety and compliance by identifying critical issues early, reducing workplace risks, and supporting regulatory adherence. Moreover, by maintaining components under optimal conditions, it helps extend the overall life of equipment, minimizing wear and avoiding premature replacements.

7.5 Future Trends and Innovations

As the manufacturing sector continues to evolve, immersive technologies are predicted to become even more integral to the process of digital transformation. Emerging innovations in these technologies are reshaping the way manufacturers design, produce, and operate. Furthermore, the growth potential of these technologies in IM offers immense possibilities to redefine not only operational efficiency but also sustainability, employee engagement, and customer-centricity.

7.5.1 Emerging Immersive Technologies

As Industry 4.0 progresses, immersive technologies are integrating more deeply with AI, ML, and cloud computing, leading to innovations that promise to transform every facet of manufacturing from design to operation to maintenance.

a. AI-Enhanced Immersion: AI is making immersive technologies more intelligent and adaptive. AI-driven simulations and predictive analytics are enabling systems to learn from user behavior and operational data, creating a more personalized and responsive experience. For example, in an MR environment, AI can analyze real-time data streams and automatically adjust the virtual representation of a machine or product based on predictive models, operational trends, or maintenance schedules.

   Moreover, in the near future, "adaptive virtual assistants" in immersive environments could feature virtual assistants powered by AI. These assistants would guide workers through tasks, provide troubleshooting suggestions, or alert them to critical issues based on real-time data analysis and ML predictions.

   Also, use of AI-powered design tools where designers could use AI algorithms in virtual environments to not only visualize but also suggest improvements and predict outcomes of various design iterations, would accelerate the design process and ensure higher quality while reducing the number of physical prototypes.

b. Haptic Feedback and Tactile Interfaces: One of the limitations of current VR and AR applications is the lack of physical interaction. While visual and auditory feedback are well-developed, the ability to feel and interact with objects in a simulated environment is still in progress. Haptic technology, which provides tactile feedback, is set to enhance immersive experiences.

   Additionally, technologies like "advanced haptic gloves" are the future immersive systems that will likely incorporate advanced haptic feedback gloves and simulate the texture, temperature, and weight of objects. This would be particularly useful in virtual prototyping and training simulations, where workers could interact with parts and machinery as if they were real.

   Moreover, "tactile interfaces" for remote operations for industries like aerospace, automotive, and oil and gas, haptic feedback can enable workers to operate machinery remotely while feeling the processes they are managing, offering new levels of precision and control.

c. 5G and Edge Computing Integration: 5G and edge computing are expected to play a pivotal role in the future of immersive technologies by providing high-speed, low-latency connectivity necessary for real-time applications.

   With the use of 5G, workers and engineers will be able to do real-time remote collaboration by engaging in highly interactive virtual environments with zero delay. These interactions could range from design reviews to complex troubleshooting tasks in real time, with immersive holographic avatars representing participants from across the globe.

Edge computing, by processing data locally on-site, will enable more immediate feedback and decision-making in immersive manufacturing systems. This will allow manufacturers to run simulations, monitor factory operations, and access digital twins even in environments where cloud computing is less reliable or time-sensitive.

d. Blockchain and Immersive Technology Integration: The integration of blockchain with immersive technologies is an emerging trend that could secure and authenticate virtual environments, digital assets, and transactions. Manufacturers could use blockchain to create trusted, decentralized digital twin networks where all interactions, changes, and maintenance histories are stored transparently. For example, in scenarios where manufacturers work with external partners (e.g., suppliers and contractors), blockchain could ensure the authenticity of digital models, designs, or proprietary data shared within the immersive environment.

7.5.2 Growth Potential

As immersive technologies are increasingly being adopted in manufacturing, their growth potential is vast, with the market to expand dramatically over the next decade. According to industry analysts, the immersive technology market for manufacturing is expected to reach billions of dollars by 2030, driven by advancements in hardware, software, and integration technologies.

a. Adoption and Market Expansion: One of the key drivers of growth is the scalability of immersive technologies. Initially, high costs limited their use to large enterprises with substantial R&D budgets. However, as the technology matures and hardware costs decrease, small and medium-sized enterprises (SMEs) are expected to be early adopters of immersive tools.
   - Affordable VR/AR Systems: With the decline in the cost of VR headsets, AR glasses, and immersive simulation platforms, more manufacturers can integrate these technologies into their operations. As affordable solutions become more accessible, SMEs will gain the opportunity to leverage immersive technology for training, prototyping, and real-time maintenance.
   - Cloud-Based Solutions: The rise of cloud platforms for immersive technologies will allow companies to access simulation, AR, and VR tools on demand, with minimal upfront investment.

This as-a-service model could significantly broaden the adoption of immersive technologies across different manufacturing sectors.

b. Digital Transformation across Sectors: As the adoption rate of immersive technologies increases, industries such as automotive, aerospace, electronics, energy, and consumer goods will harness the power of these systems to improve operations, create new products, and enhance customer engagement.
   - Automotive Manufacturing: Immersive technologies are already transforming vehicle design, assembly, and testing. With advances in simulation and digital twins, manufacturers will be able to predict and adjust for wear and tear, road conditions, and manufacturing defects before production begins.
   - Aerospace: The aerospace industry is set to adopt immersive training platforms for pilots, engineers, and ground staff. These platforms will simulate complex scenarios ranging from aircraft system failures to emergency procedures, ensuring that aerospace operations are safe and efficient.
c. Sustainability and Efficiency Gains: As the world continues to face pressure to reduce carbon emissions and resource waste, immersive technologies offer a unique way to reduce material waste and energy consumption.
   - Resource Optimization: Through immersive simulations, manufacturers can design more efficient production processes, reduce energy usage, and minimize material waste.
   - Sustainable Prototyping: Virtual prototyping and simulation will drastically reduce the need for physical models, thus conserving raw materials and reducing environmental impact.
d. Workforce Transformation: As the demand for more skilled workers increases, manufacturers will rely on AR, VR, and MR to accelerate training and increase productivity.
   - Remote Workforces: As companies continue to embrace remote work, immersive technologies will bridge the gap between on-site and off-site employees, fostering collaboration in real-time across geographies and time zones.
e. Next-Generation Supply Chains: With the integration of immersive technologies into supply chain management, manufacturers will gain more insight and control over their global operations. Using AR overlays and digital twins, companies can track materials, shipments, and products in real time, enhancing visibility and efficiency across the entire supply chain.

7.6 Highlights

- Immersive Technologies such as VR, AR, and MR are transforming IM by enabling interactive, intuitive, and data-rich environments for operators and engineers.
- VR provides a fully immersive, computer-generated environment that is particularly valuable in product design, prototyping, and virtual training simulations, minimizing the need for physical prototypes and accelerating design validation cycles.
- AR enhances the real-world environment with digital overlays, supporting maintenance operations, real-time guidance, and training by visualizing hidden components, workflows, or error-prone areas directly in the operator's field of view.
- MR combines physical and virtual elements for dynamic interaction. In IM, MR supports collaborative decision-making, remote troubleshooting, and adaptive automation, creating shared spaces where digital tools and humans interact seamlessly.
- Immersive technologies are being increasingly integrated into IM applications, such as
    - Product design and prototyping—accelerating iterations and improving design accuracy.
    - Training and skill development—enhancing safety and effectiveness with realistic, scenario-based learning.
    - Smart factory operations—providing contextual data in real-time to improve productivity.
    - Quality control and inspection—enabling better visualization of defects and deviations through advanced imaging and overlay systems.
    - Simulation and Digital Twins act as complementary tools, replicating physical manufacturing environments digitally for process optimization, predictive maintenance, and risk-free experimentation.

    The future trends point to a convergence of AI, immersive technologies, and cyber-physical systems, leading to fully adaptive, user-centric, and decentralized manufacturing ecosystems.

Exercises

P7.1 Differentiate between VR, AR, and MR with suitable examples from manufacturing.

P7.2 What are the key benefits of using immersive technologies in IM?

P7.3 How does Mixed Reality contribute to collaborative manufacturing environments?

P7.4 Design a training module using VR for machine operators in a smart manufacturing facility. What hardware and software would you use?

P7.5 Explain how AR can be applied to assist in real-time maintenance and repair tasks on complex machinery.

8

Product Life Cycle Management, Big Data and Cloud Manufacturing

8.1 Introduction

In the age of Industry 4.0, manufacturing systems are undergoing a digital transformation characterized by connectivity, intelligence, and adaptability. Central to this transformation are three interrelated pillars: Product Life Cycle Management (PLM), Big Data Analytics, and Cloud Manufacturing (CMfg). These technologies are redefining how products are designed, produced, and serviced and enabling organizations to achieve higher levels of efficiency, customization, and responsiveness in global markets.

PLM is a strategic approach that manages the complete journey of a product, ranging from initial concept, through design and manufacturing, to service and eventual disposal. PLM ensures the integrity and accessibility of product-related data across all life cycle stages, fostering collaboration among diverse stakeholders, including designers, engineers, suppliers, and customers. It enables traceability, accelerates innovation cycles, and supports regulatory compliance in highly complex and data-driven environments.

Complementing PLM is the power of Big Data Analytics, which allows manufacturers to harness large volumes of heterogeneous data generated across the product life cycle. This includes data from design revisions, shop floor sensors, customer feedback, and maintenance logs. Advanced analytics techniques, such as machine learning, predictive modeling, and statistical inference, provide actionable insights for improving product quality, optimizing resources, reducing downtime, and enhancing decision-making processes.

CMfg, on the other hand, offers a paradigm shift in how manufacturing resources and capabilities are provisioned and consumed. CMfg enables flexible access to software, data, and equipment as services by integrating cloud computing with manufacturing systems. It supports distributed collaboration, dynamic resource allocation, and real-time scalability, unlocking the potential for "manufacturing-as-a-service" models and virtual enterprise ecosystems.

The convergence of PLM, Big Data, and CMfg represents more than a technological upgrade. It reflects a systemic transformation toward agile,

DOI: 10.1201/9781003485186-10

intelligent, and sustainable manufacturing. This chapter explores the foundational concepts, integration frameworks, and practical applications of these pillars.

8.2 Product Life Cycle Management (PLM)

PLM systems are built to unify product development activities, enhance design collaboration, and streamline engineering change management. They are especially critical in sectors like automotive, aerospace, and medical devices, where design complexity, compliance requirements, and product longevity demand robust control of versioning, documentation, and communication.

8.2.1 Components of PLM

8.2.1.1 Product Data Management (PDM)

Managing the growing volume and complexity of "design data" presents significant challenges in modern product development environments. Product Data Management (PDM) addresses these challenges by offering a centralized system for organizing, accessing, and controlling all product-related information throughout its life cycle. The need for PDM stems from the demand for efficient information sharing, version control, and data security. It ensures that design teams have access to the latest data, safeguards against data loss, and supports standardization across product structures and BOM. This is particularly vital for managing multiple product variants, tracking project status, and implementing timely and informed engineering changes. PDM also supports compliance with international standards by promoting structured and traceable workflows. It is estimated that in companies without a strong PDM system, engineers spend 25% more time on ineffective data management chores.[1]

PDM enables seamless integration of various data types, such as CAD models, engineering documents, simulation results, manufacturing plans, and multimedia files, into a single coherent platform. More than a storage solution, PDM systems enforce data integrity, regulate user access, and establish relationships between data items. These platforms also include notification systems and workflow engines that alert stakeholders to design updates and facilitate real-time collaboration. Through these capabilities, PDM ensures that critical engineering information is delivered efficiently to downstream systems such as CAE, CAPP, CAM, ERP, and manufacturing execution platforms.

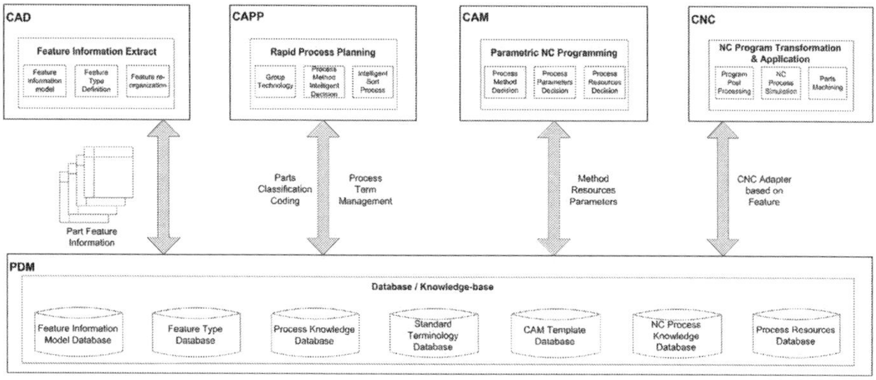

FIGURE 8.1
PDM and its applications in product development.

In practice, PDM empowers multidisciplinary teams by providing visibility, traceability, and control over the evolution of a product. Engineers, planners, and managers can coordinate across departments while ensuring data revisions are captured and documented. For example, when introducing a new product variant, a PDM system checks for consistency with original specifications and notifies relevant teams of potential impacts. Such functionality accelerates the engineering change process and minimizes costly errors. Figure 8.1 shows PDM and its applications in product development.

PDM offers several benefits for the speedy execution of product development projects. For instance:

- It helps access to drawings by engineers in minutes.
- Managers know the number of drawings to be released for fabrication.
- Manufacturing engineers know exactly when to start fabrication.
- Two designers cannot simultaneously edit the same drawing.
- Two drawings cannot have the same drawing numbers.
- All information from drawing-release to revision details pass to the primary project management system that tracks overall project status for all the company activities.
- The PDM system keeps a complete history of all the modifications, along with when the modifications took place, and keeps all previous versions of data for a defined period (given by the company).
- Subcontractors can have direct access to the PDM database, and it is ensured that no subcontractor can access data associated with the subsets of projects they are not involved with.

Popular PLM software that includes PDM modules are Siemens Teamcenter, PTC Windchill, Solidworks PDM, Autodesk Vault, and OpenBOM.

8.2.1.2 Configuration Management

Configuration management is a crucial component of PLM that provides organized control over product variations, alternatives, and revisions across several product lines. It ensures that every configuration is consistent, traceable, and in line with business needs by methodically managing modifications to product structures, design specifications, and documentation. Configuration management allows companies to respond to market needs more quickly, handle customization effectively, and support regulatory compliance while maintaining product integrity over time.

For example, in automotive manufacturing, companies typically produce various models of the same vehicle, each featuring different specifications such as engine types, trim levels, and optional features like sunroofs, navigation systems, or hybrid powertrains. Configuration management plays a critical role in managing these variations. It ensures that each specific product configuration, comprising distinct parts and assemblies, is accurately captured, monitored, and updated over its entire life cycle. For example, when an update is made to the vehicle's multimedia system, the change is applied only to the versions equipped with that feature.

8.2.1.3 Process Management

Process Management within PLM refers to the structured coordination of activities, approvals, and deliverables throughout a product's development journey. It manages workflows by defining how tasks are initiated, reviewed, and completed, ensuring that each step in the development process follows a clear and traceable path. Organizations can automate approval chains, monitor project timelines, assign responsibilities, and track task progress across departments. This improves visibility and accountability and reduces delays caused by manual follow-ups or miscommunication.

For example, when a design change is proposed, the system automatically routes the request through the appropriate engineering, quality, and compliance stakeholders for review and sign-off.

8.2.1.4 Collaboration Tools

One of the key strengths of modern PLM systems is their ability to support collaboration among teams that are distributed across different locations and disciplines. Collaboration tools embedded within PLM platforms allow engineers, designers, supply chain managers, and quality professionals to work on the same product data in real time, regardless of where they are based.

These tools facilitate shared access to 3D models, design files, drawings, and change requests, while maintaining strict control over versioning and user permissions. By breaking down silos and improving communication, collaboration tools in PLM enhance productivity, reduce design errors, and support faster decision-making.

For instance, a design update made by an engineer in one location can be instantly reviewed and commented on by a colleague in another region. This shortens feedback cycles and fosters better alignment between departments such as R&D, manufacturing, and procurement.

8.2.1.5 Integration with ERP and MES

To fully understand the benefits of digitalization in manufacturing, the PLM system must communicate effectively with other enterprise platforms like ERP and MES. PLM-ERP integration ensures that product design data flow seamlessly into procurement and production planning activities, such as part numbers, BOMs, and engineering changes. Similarly, linking PLM with MES bridges the gap between design and manufacturing departments, allowing real-time synchronization of manufacturing instructions, process parameters, and quality requirements. Moreover, this integration eliminates the need for manual data transfer, reduces the risk of discrepancies, and ensures that the latest approved design is always used in production.

For example, when an engineering change is finalized in the PLM system, it can automatically update associated manufacturing orders and inventory records in the ERP and MES platforms. The result is a more agile, error-resistant, and synchronized manufacturing process.

8.2.2 Example of Application of PLM in the Automotive Sector

Today, leading OEMs must coordinate design, development, and production across continents. A compelling example of PLM can be seen in how global automotive companies utilize platforms like Siemens Teamcenter or PTC Windchill to manage the complete digital thread of a vehicle's life cycle.

Consider a scenario where a European car manufacturer designs a new mid-size hybrid sport utility vehicle (SUV). The project spans multiple geographies. The core design team is based in Munich, Germany; the supply chain operations are handled from a regional hub in Mexico, while final assembly is carried out at a state-of-the-art manufacturing plant in Wuhan, China. Despite the geographical spread, all stakeholders access a unified PLM platform that integrates engineering drawings, CAD models, BOM, supplier information, and real-time compliance documentation.

When a quality issue is reported, like a recurring fault in the infotainment system's touch display, the engineering team in Germany raises an

Engineering Change Request (ECR) within the PLM system. This request is automatically routed to relevant teams via predefined approval workflows. Quality engineers in China assess production impacts, procurement specialists in Mexico re-evaluate supplier timelines, and project managers ensure that design updates are rolled out across all digital and physical domains. Every stakeholder sees a live, traceable history of the issue and its resolution, ensuring transparency and audit readiness.

Moreover, the PLM system supports multiple vehicle configurations, like petrol-only variants for selected markets and fully electric versions. Each variant's BOM and associated documentation are managed intelligently so that changes to one configuration do not affect others. As a result, design integrity, regulatory compliance, and time-to-market are preserved, showing that the integrated digital workflow of PLM doesn't just store information; it controls product innovation across the entire value chain.

8.3 Big Data

Big data refers to vast and complex datasets that cannot be analyzed with conventional database systems and traditional data processing tools. These datasets originate from various sources within manufacturing facilities, such as machines, sensors, supply chains, and customer feedback. With such vast and complex data coming in, capture, storage, curation, transfer, analysis, and visualization challenges emerge.

In IM, big data analytics plays a transformative role by enabling the collection and interpretation of this high-volume data to drive operational efficiency, improve product quality, and support real-time decision-making. The value of big data lies in its ability to explore meaningful patterns and correlations within a unified dataset, which can be leveraged to identify trends, monitor quality, optimize logistics, prevent equipment failures, and respond proactively to customer needs.

8.3.1 Characteristics Associated with Big Data

Figure 8.2 shows the five Vs associated with big data:

- Volume shows that a huge amount of data is generated daily from diverse sources such as sensors, machines, and enterprise systems. Examples include terabytes of data from shop floor sensors, supply chains, and customer usage logs.
- Velocity emphasizes the rapid pace at which IM systems create, transmit, and process data. For example, the data from IoT-enabled

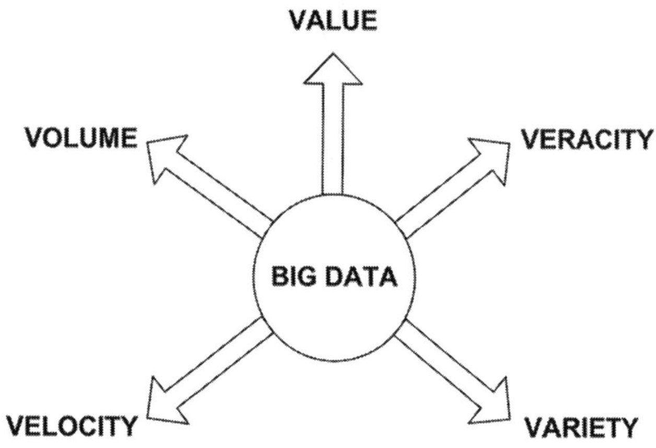

FIGURE 8.2
Five Vs associated with big data.

devices must be captured and acted upon in real time to support responsive decision-making.

- Variety signifies the different forms the data can take, from being structured (e.g., spreadsheets), semi-structured (e.g., XML), to unstructured (e.g., videos, images, and text). For example, the data might include everything from machine logs to supplier invoices and product reviews.
- Veracity addresses the quality, reliability, and uncertainty of data. In industrial environments, noise in sensor readings, missing entries, or inconsistent formats can compromise analysis if not handled efficiently.
- Value refers to extracting meaningful and actionable insights from the massive datasets. For instance, in manufacturing, this means identifying defect patterns, predicting equipment failure, or optimizing energy usage.

With the passage of mine, a few more Vs of big data have also emerged. "Validity" refers to the accuracy and correctness of data, ensuring that insights drawn are reliable and actionable. "Variability" shows the data's unpredictable nature or dynamic behavior, which may fluctuate in patterns or structure over time. "Volatility" denotes how long data remains relevant or useful, emphasizing the importance of timely analysis before its value diminishes. "Vulnerability" highlights the security risks associated with big data, such as potential breaches or cyberattacks, making data protection a critical concern. Finally, "Visualization" involves translating complex datasets into visual formats, enabling better interpretation, faster decision-making, and broader accessibility for diverse stakeholders.

8.3.2 Big Data Analytics

From Section 8.3.1, it is evident that raw data can have many associated characteristics, and its timely inspection, organization, and interpretation is essential to extract useful information that can support decision-making. With the rapid digitalization of manufacturing environments, data analysis is increasingly getting automated through mechanical systems and algorithms, enabling continuous analysis at scale and speed because traditional data analysis can't handle the complexity of data these days.

Traditional data analysis primarily employs statistical methods to examine large and complex datasets. It aims to uncover hidden patterns, trends, and correlations that are not readily visible. By exploring and elaborating on this hidden data, traditional analysis has helped maximize the value embedded within datasets for years and has served as a tool for customer demand forecasting and market trend predictions. The modern shift, however, is toward more dynamic and automated forms of analysis, where advanced tools process real-time data for immediate insights that influence business operations and strategic decisions.

Now, the data can broadly be categorized into two types: quantitative and qualitative. Quantitative data includes numerical values that can be measured or compared on a scale, such as demographic statistics, survey scores, workforce attendance records, and standardized test results. This form of data is suitable for statistical analysis and visualization. Qualitative data, on the other hand, is descriptive and non-numerical, often derived from interviews, observations, or open-ended survey responses. It captures subjective aspects such as user experience, opinions, or design preferences. Both data types play essential roles in IM, where a combination of numerical metrics and contextual insights is necessary for a holistic understanding of operations, performance, and user feedback.

8.3.2.1 Level of Analysis

Figure 8.3 shows the different levels of analysis.

Descriptive analytics is the foundation of data analysis and is found at the bottom. It focuses on summarizing historical data to understand what has occurred in the past. Using basic statistical techniques like averages, percentages, and standard deviations, it organizes and presents information through charts, graphs, or dashboards. This type of analysis doesn't predict or explain outcomes but prepares data for deeper analysis. For example, it might highlight monthly sales trends, changes in production volume, the amount of time a few machines are not functional, defects identified per batch, etc. AI and ML incorporation at this stage is rarely seen, and the level of advancement of the industry is also "low." Almost all organizations use descriptive analytics.

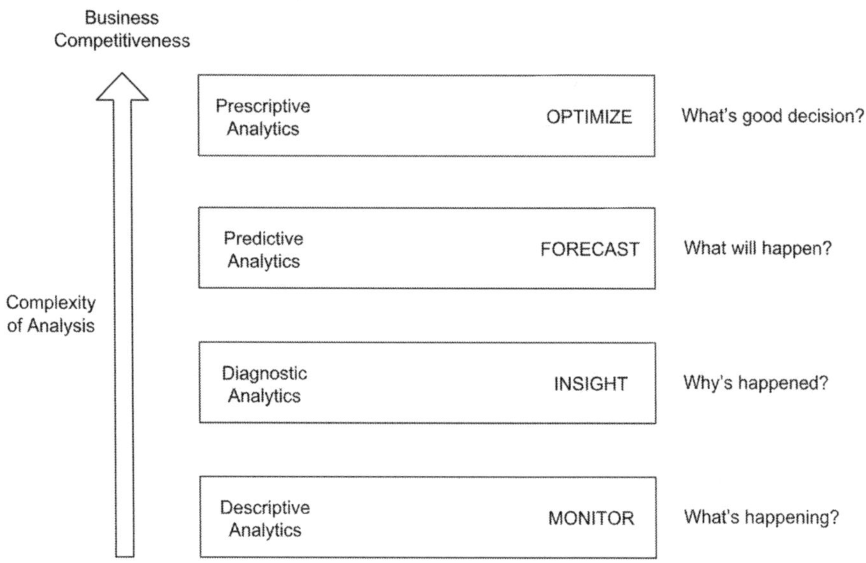

FIGURE 8.3
Levels of analysis.

Building on descriptive insights, diagnostic analytics (one level up) explores the reasons behind observed outcomes. It uses techniques like data mining, correlation, and sensitivity analysis to investigate the root causes. This stage is especially useful for finding out which input parameters impact a certain outcome. For example, which customer segment drove a sales increase, or what caused delays in a supply chain? Diagnostic tools help teams make informed decisions by offering a clearer understanding of interrelationships within data. This type of analytics incorporates AI and ML, sometimes with the level of advancement of the industry as "medium." Many organizations use diagnostic analytics.

Further up, predictive analytics leverages statistical models and machine learning algorithms to forecast future outcomes and answer the question "What will happen?" This type of analysis enables proactive decision-making by identifying trends and estimating potential risks or opportunities. AI and ML are usually incorporated in predictive analytics, and the industry's level of advancement is "high." It is used by a smaller yet growing number of industries.

At the top of Figure 8.3 is prescriptive analytics, which suggests the optimal course of action by simulating various decision scenarios and outcomes. It addresses the question "What's the best decision?" and is used for strategic optimization in areas like inventory planning, maintenance scheduling, and process control. AI and ML are always incorporated in such type of analytics, and the level of advancement of industries incorporating them is "very high." Prescriptive analytics is not yet widespread in industries.

Overall, from bottom to top, the complexity of methods and the business competitiveness increase significantly, making advanced analytics a cornerstone of IM.

8.3.2.2 How Big Data Analysis Works?

Figure 8.4 shows the flow of information in the analysis of big data. An example of health sciences is used.[2]

- Formulate Question: The process starts with identifying meaningful and solvable questions, such as predicting cancer susceptibility or monitoring a machine's health condition. Since not all problems require big data, selecting a relevant, data-driven question is key.
- Data Collection: Once the question is formulated, data is gathered using various sources like mobile phones, social media platforms, omics profiling technologies, and hospital records. These sources reflect IoT and the growing ecosystem of smart health devices.
- Data Storage and Transfer: The collected data, which is often in terabytes and heterogeneous, needs secure and efficient storage.

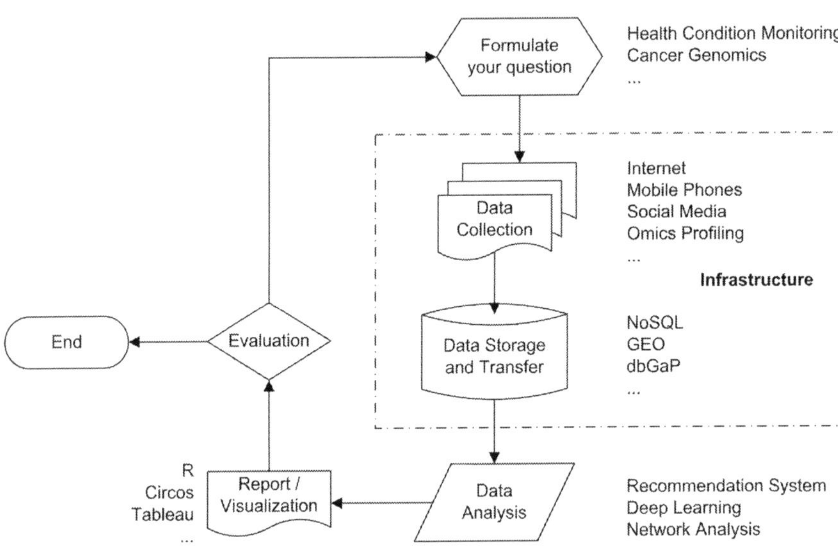

FIGURE 8.4
Flow of information in big data analysis.

Technologies like NoSQL databases, GEO, or dbGaP are employed for this purpose. Data transfer protocols like IBM Aspera may be used to handle large volumes securely and at high speed.

- Data Analysis: Advanced computational techniques are applied to make sense of the data at this stage. Tools such as deep learning, network analysis, and recommendation systems help extract patterns, correlations, and predictions from complex datasets.
- Report/Visualization: The results are translated into visual and digestible formats using platforms like R, Circos, or Tableau. These help stakeholders understand the analysis quickly and effectively.
- Evaluation: This final stage checks whether the analysis has answered the initial question. If not, the cycle restarts with refining the question or improving the data or methods used.

8.3 Big Data Applications in Manufacturing

- Predictive Maintenance: By analyzing data from sensors embedded in equipment, manufacturers can predict when machines are likely to fail. For example, a turbine manufacturer might monitor temperature, vibration, and noise data to detect bearing wear before failure occurs, reducing unplanned downtime.
- Demand Forecasting: Big data models can analyze past sales, market trends, and even weather patterns to predict demand, helping optimize inventory and reduce waste accurately.
- Quality Control: Statistical process control combined with ML allows manufacturers to detect deviations from quality benchmarks in real time. In semiconductor fabrication, image-based inspection data can be analyzed to catch microscopic defects early in the process.
- Process Optimization: Real-time analytics can optimize energy consumption, material usage, and cycle times. For example, a smart factory might use historical production data to identify the most energy-efficient combination of machine settings to manufacture a particular batch in a reconfigurable manufacturing system (RMS).
- Supply Chain Visibility: When combined with big data analytics, IoT-enabled logistics data provides a holistic view of the supply chain, improving responsiveness to disruptions and customer demands.

8.4 Cloud Manufacturing

CMfg is a new service-oriented networked manufacturing paradigm that aims to solve more complex manufacturing problems and carry out larger-scale collaborative manufacturing. It transforms manufacturing resources and capabilities into manufacturing services through network, cloud computing, service computing, and manufacturing-enabling technologies (see Figure 8.5).

A comprehensive CMfg system leverages advanced technologies such as embedded sensors, IoT, cloud computing, and virtualization to enable real-time monitoring, resource sharing, and scalable manufacturing services. As a network-based model, CMfg allows users to access and deliver on-demand services across the product life cycle, with individuals dynamically acting as service providers or requesters. Cloud platforms coordinate distributed production facilities and are managed by operators to ensure seamless, flexible service delivery. CMfg also supports customized and agile production by integrating cyber-physical systems, AM, and data-driven tools like simulation and optimization. The system's effectiveness relies on collaboration among three key stakeholders: operators, providers, and users, which

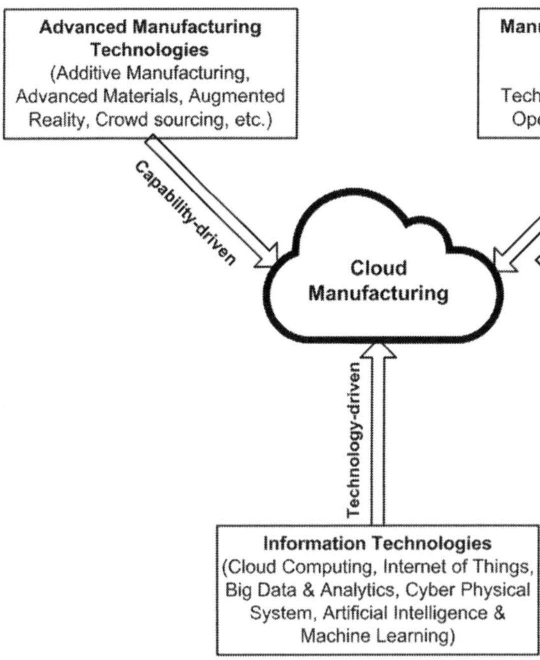

FIGURE 8.5
Concept of cloud manufacturing.

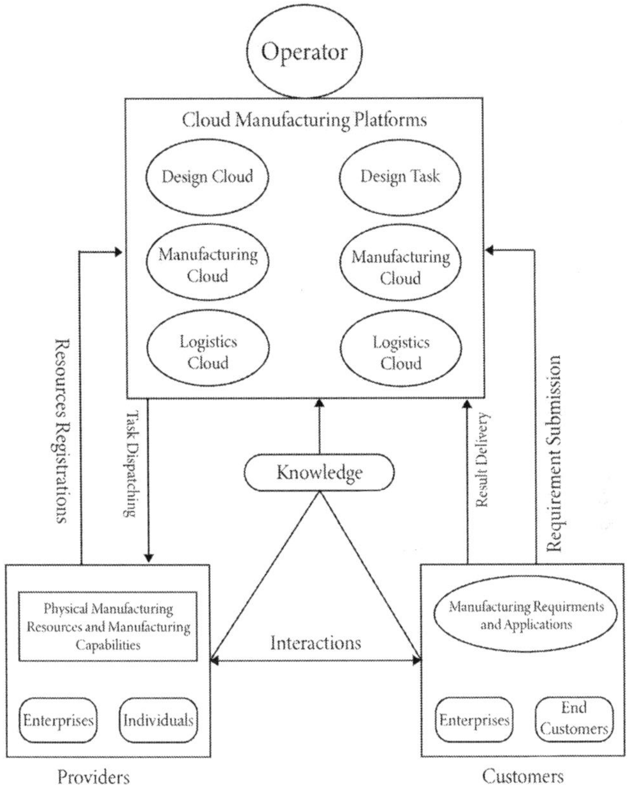

FIGURE 8.6
Operational model of cloud manufacturing.

ensure efficient, collaborative, and adaptive manufacturing across the value chain (see Figure 8.6).[3]

8.4.1 Characteristics of Cloud Manufacturing

CMfg encompasses distinctive features enabling it to support modern, efficient, and responsive manufacturing systems. These characteristics include adaptability and scalability, on-demand services, knowledge-driven operations, manufacturing as a service, and multi-tenancy.

- Adaptability and Scalability: Adaptability refers to the system's ability to respond to changes in manufacturing conditions, while scalability allows the system to handle increased production demand without compromising performance. Real-time data from embedded sensors and IoT devices enables continuous monitoring and adjustment of production setups.

Example: A smart factory producing consumer electronics can automatically reconfigure its assembly line based on changes in product design specifications or demand levels.

- On-Demand Services: On-demand manufacturing allows customers to request products tailored to their needs. This includes customization in terms of size, material, color, or function.
  Example: A 3D printing service produces personalized orthopedic insoles based on digital foot scans provided by the customer.
- Knowledge-Driven Operations: CMfg analyses large volumes of operational data to optimize manufacturing processes. This includes process parameters, quality control data, and maintenance logs. The insights gained support informed decision-making and continuous improvement.
  Example: A metal casting plant using historical temperature and cooling data predicts defects and adjusts casting parameters automatically.
- Manufacturing as a Service (MaaS): CMfg supports outsourcing specific manufacturing tasks or components to third-party providers through the cloud platform. This allows companies to access specialized capabilities without owning the physical infrastructure.
  Example: An automotive company subcontracts the production of a precision-machined gear component to a cloud-based supplier with advanced CNC machining capabilities.
- Multi-Tenancy: Multi-tenancy allows multiple users or organizations to share the same cloud infrastructure while maintaining data and process isolation. It enables manufacturers to utilize excess capacity from other firms, improving resource utilization and fostering collaboration.
  Example: A small-scale medical device startup accesses sterilization services from a larger pharmaceutical manufacturer through a shared CM platform.

8.4.2 Application Example of Cloud Manufacturing in Aerospace Production[3]

CA Corporation is a large, multi-site enterprise in China, primarily engaged in designing and manufacturing aerospace products. CA operates in a highly complex and multidisciplinary environment with over 600 self-financing subsidiary companies distributed across various regions. Each aerospace product model typically involves significant financial investment, lengthy development timelines, and the integration of numerous technical disciplines across multiple collaborators. Despite its size and technical capabilities, CA has faced challenges including redundant infrastructure, communication gaps, and inefficient collaboration among subsidiaries. To address these

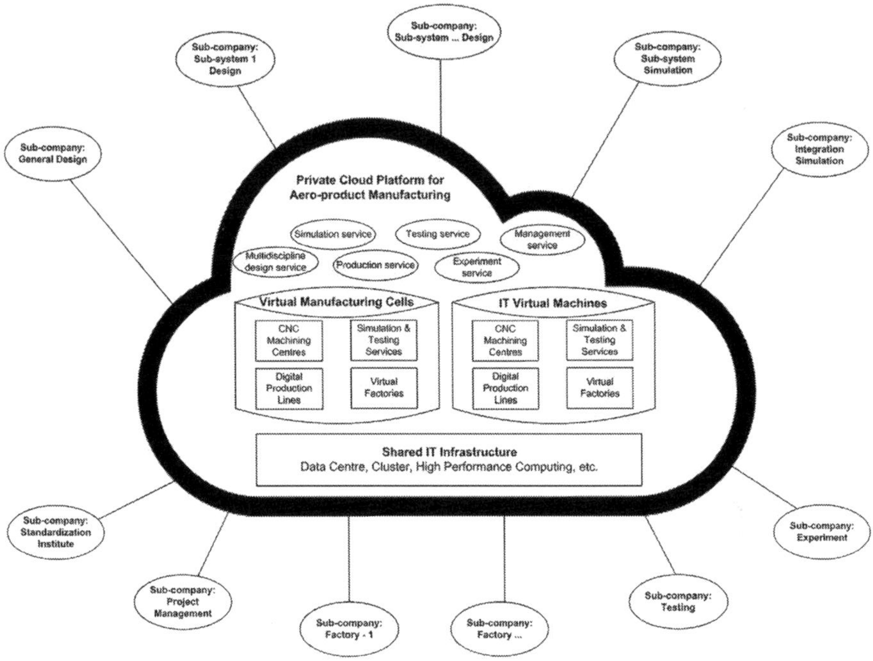

FIGURE 8.7
A private cloud manufacturing system in a super-division of CA corporation.

inefficiencies, CA adopted the CMfg model as a transformative approach to enterprise-wide resource integration and collaboration.

One key implementation of this concept is shown in Figure 8.7, which presents a private cloud platform for aero-product manufacturing managed by a central cloud operation center under CA's super-division. This super-division includes dozens of sub-companies, each responsible for different product life cycle stages, from general and subsystem design to simulation, production, assembly, testing, experimentation, and project management.

To support this collaborative framework, CA developed a robust shared IT infrastructure and data center, forming the technological backbone of its cloud platform. Its primary goal is to enable seamless integration of digital resources from sub-companies into a unified virtual resource pool. IT resources such as servers, storage systems, operating systems, CAx software, and management tools are encapsulated into virtual machines (VMs) for cloud-based use.

However, integrating hard resources like traditional manufacturing equipment into the cloud remains challenging, particularly when devices lack digital interfaces. Even advanced IoT methods (e.g., RFID sensors) may be ineffective without digital readiness. Therefore, a successful CMfg system relies heavily on enterprise digitalization. In the developed CMfg platform, only digital-ready assets such as CNC machining centers,

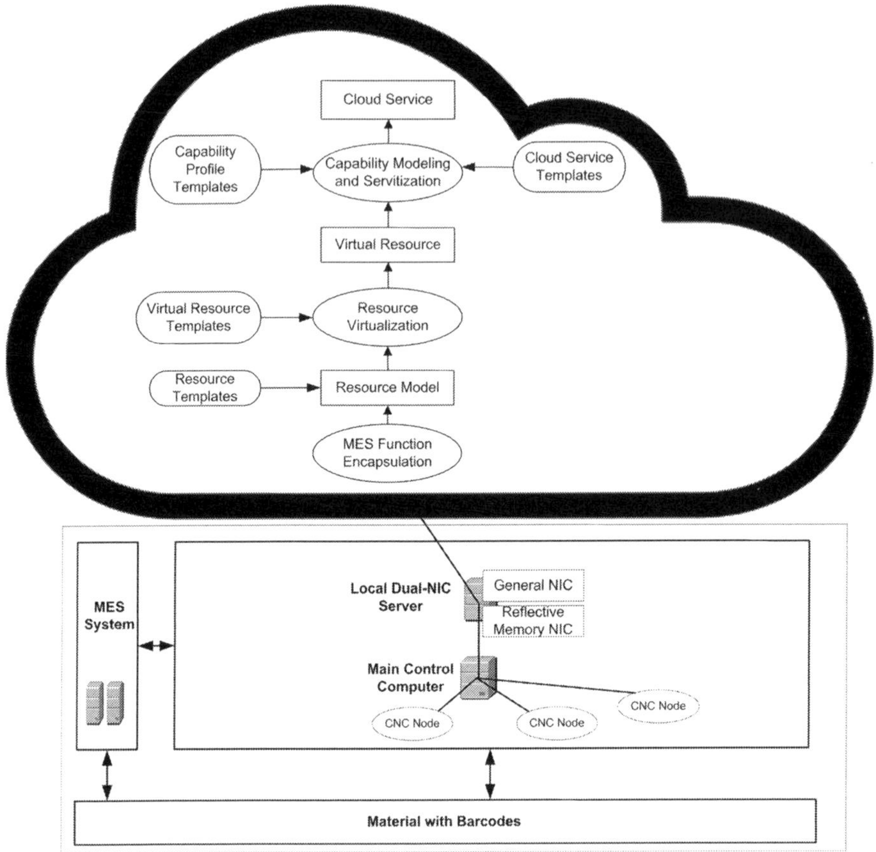

FIGURE 8.8
An example of hard resource connection, virtualization and servitization.

simulation/testing devices, and automated production lines are included in the virtual manufacturing cells.

Moreover, as shown in Figure 8.8, the CNC machines are connected to a main-control computer linked to a dual-network interface card (NIC) server; one for general networking and another (reflective memory NIC) for real-time, low-latency communication. To maintain cost efficiency, production materials are tracked using barcodes rather than RFID tags.

Prior to cloud integration, CA utilized a MES to manage production logistics. The MES plays a pivotal role in virtualizing hard assets by exposing interfaces that can be wrapped into web services. These interfaces, once virtualized, allow CNC nodes to be represented as logical machines, with production capabilities that can be profiled using metrics like quality of service (QoS), technical expertise, and historical performance. Based on these profiles, capabilities are published on the cloud as services, such as OnCloud remote monitoring or OffCloud machining.

8.5 Highlights

The highlights of this chapter are as follows:

a. Integration of Key Industry 4.0 Technologies: Demonstrates how PLM, Big Data, and CMfg collectively enable IM and digital transformation.
b. A comprehensive PLM Framework covers essential PLM components such as PDM, configuration management, process workflows, collaboration tools, and integration with ERP/MES systems.
c. Real-World PLM Application illustrates how global automotive companies use PLM platforms like Siemens Teamcenter and PTC Windchill to manage the digital thread of vehicle development across continents.
d. Understanding Big Data in Manufacturing: Explains Big Data's characteristics (Volume, Variety, Velocity, Veracity, Value) and emerging dimensions like Validity, Vulnerability, and Visualization.
e. Multi-Level Analytics: Describes descriptive, diagnostic, predictive, and prescriptive analytics with industry use cases and AI/ML integration levels.
f. Big Data in Action: Showcases applications such as predictive maintenance, demand forecasting, quality control, process optimization, and supply chain visibility.
g. Cloud Manufacturing Architecture: Details the structure and operation of CMfg systems, including adaptability, on-demand services, knowledge integration, MaaS, and multi-tenancy.
h. Case Study of CMfg: Presents a large-scale deployment of CMfg in the aerospace sector involving 600+ sub-companies, highlighting resource virtualization and MES integration.

Exercises

P8.1 Explain the role of PDM in reducing errors during engineering change processes.

P8.2 What is configuration management in the context of PLM, and how does it support the handling of product variants in industries such as automotive or electronics manufacturing?

P8.3 Describe the key dimensions (Vs) of Big Data and explain how each one impacts decision-making in an IM environment.

P8.4 Compare and contrast the four layers of analytics: descriptive, diagnostic, predictive, and prescriptive, used in big data applications. Provide manufacturing-specific examples for each.

P8.5 **Application-based problem**

A smart manufacturing firm is looking to implement a cloud manufacturing system. It currently has three CNC machines, one 3D printer, and two testing units. These machines are digitally enabled but operate independently. The company struggles with delayed order delivery, underutilized machines, and a lack of remote monitoring capabilities.

Tasks:

a. Propose how these resources could be virtualized and offered as services.

b. Identify which cloud manufacturing features (e.g., on-demand, scalability) would benefit this company.

c. Suggest how a MES could connect machines to the cloud.

Notes

1 What Is Product Data Management? *PDM Software*. PTC. (2024, November 19). https://www.ptc.com/en/technologies/plm/product-data-management

2 Huang, T., Tan, L., Fang, X., An, P., Min, J., & Wang, F. (2015). Promises and challenges of big data computing in health sciences. *Big Data Research*, 2(1), 2–11. http://dx.doi.org/10.1016/j.bdr.2015.02.002

3 Liu, Y., Wang, L., Wang, X. V., Xu, X., & Jiang, P. (2019). Cloud manufacturing: Key issues and future perspectives. *International Journal of Computer Integrated Manufacturing*, *32* (9), 858–874.

9

Converging Technologies for Smart, Circular, and Energy-Efficient Manufacturing

9.1 Machine/Computer Vision[1]

9.1.1 Introduction

Computer vision is a specialized branch of AI and computer science that enables machines to interpret and derive insights from visual inputs. This capability is applied in industrial settings through machine vision, systems designed to replicate human visual perception for complex operations such as inspection, component tracking, and automated analysis. Computer vision becomes a central component in achieving operational efficiency and quality assurance when integrated with IM infrastructures, known for their connectivity, adaptability, and data-driven control.

9.1.2 Enabling Technologies in Computer Vision

The foundation of computer vision systems is built upon a range of interconnected technologies:

- Image Acquisition: Visual data is collected using different types of sensors and cameras, including RGB, infrared (IR), and depth-sensing devices, forming the basis for further processing.
- Preprocessing Methods: Raw images often require enhancement through techniques such as contrast adjustment, edge enhancement, and noise reduction to prepare them for analysis.
- Extraction of Features: Key visual features like edges, textures, contours, and shapes are extracted using algorithms such as Scale-Invariant Feature Transform (SIFT), Histogram of Oriented Gradients (HOG), and Oriented FAST and Rotated BRIEF (ORB).
- Object Detection and Classification: Advanced deep learning techniques, including CNNs, You Only Look Once (YOLO), Faster R-CNN, and Single Shot MultiBox Detector (SSD), enable the identification and classification of objects in real time.

DOI: 10.1201/9781003485186-11

- Image Segmentation: Breaking down images into defined regions allows systems to isolate and analyze areas of interest crucial for high-precision inspection tasks.
- 3D Vision and Depth Perception: Using approaches like stereo vision, structured light, and simultaneous localization and mapping (SLAM), machines gain depth perception, enabling spatial reasoning and guidance.
- Integration with Deep Learning (DL): Vision systems increasingly rely on deep neural networks, vision transformers, and Generative Adversarial Networks (GANs) to handle complex operations like pose estimation, semantic segmentation, and anomaly detection, allowing high-level interpretation of complex visual patterns.
- Sensor Fusion and Edge Computing: Integrating multiple sensors with real-time edge computing platforms enables fast and decentralized decision-making on the production floor.

9.1.3 Industrial Applications of Computer Vision in Intelligent Manufacturing

In IM environments, computer vision facilitates:

- Automated Quality Control: Systems equipped with vision tools can identify product defects such as misalignments, scratches, or color variations, significantly improving inspection efficiency.
- Condition Monitoring and Maintenance: Visual monitoring of machine parts for signs of wear and anomalies supports condition-based maintenance strategies.
- Robotic Operations and Assembly: Vision-enabled robots adjust to dynamic settings, allowing them to carry out precision assembly and material handling, particularly in low-volume, high-variation production.
- Digital Twin Development: Live visual data from production lines enriches digital twin models, aiding in simulation, monitoring, and real-time optimization.
- Monitoring Additive Manufacturing (AM): In 3D printing processes, camera systems monitor each layer, ensuring precise material placement and identifying early signs of build failure.

In practice, automotive plants use vision systems to verify welds and paint quality, while electronics manufacturers employ them to detect flaws in micro-scale circuitry.

9.1.4 Challenges in Implementation

Implementing computer vision solutions in manufacturing environments involves several obstacles:

- Visual Variability: Factors like inconsistent lighting, glare, or partial occlusion can reduce system accuracy.
- Processing Demands: High-resolution, real-time image processing requires considerable computational resources, especially when deployed at scale.
- Generalization of Models: AI models trained under specific conditions may struggle to maintain performance when faced with new product types or environmental changes.
- Legacy Integration Issues: Merging vision systems with existing production hardware and software infrastructures can be complex and costly.
- Talent Shortage: There remains a notable skills gap in the workforce when it comes to deploying and managing AI-based vision systems in industrial contexts.

9.1.5 Future Directions

To address these challenges and fully integrate computer vision into IM workflows, several strategies can be pursued:

- Deployment of Edge and Hybrid AI Architectures: Processing data closer to the source using edge devices reduces latency and enables faster decisions.
- Standards and Interoperability Frameworks: Establishing standard protocols for image formats, communication, and system interfaces will aid system scalability.
- Cross-Functional Training Initiatives: Upskilling engineering teams in machine vision and AI ensures long-term success and adaptability.
- Leveraging Digital Twins: Coupling vision data with virtual simulations allows manufacturers to test and optimize processes in advance.

Conclusively, as the visual intelligence layer of manufacturing systems, computer vision enhances automation, quality assurance, and adaptive decision-making. With advancements in deep learning, edge computing, and sensor technologies, their applications are becoming broader and more impactful. Addressing current limitations through investment in standards, training, and infrastructure will be essential to fully realizing its potential in IM systems of the future.

9.2 Additive Manufacturing

9.2.1 Introduction

AM, commonly known as 3D printing, has emerged as a foundational technology in the landscape of Industry 4.0. Characterized by its capacity to construct intricate geometries layer-by-layer from digital models, AM signifies a departure from traditional subtractive and formative manufacturing paradigms. This transformative capability aligns with the broader objectives of Industry 4.0, including digitalization, sustainability, and enhanced responsiveness to market needs.

In parallel, IM has gained prominence by embedding real-time data analytics, CPS, AI, and cloud computing into manufacturing operations. The fusion of AM and IM not only boosts production agility and efficiency but also fosters sustainable, decentralized manufacturing ecosystems. Together, they represent the building blocks of the FoF.

9.2.2 Additive Manufacturing: Core Principles and Process Flow

According to ASTM, AM is a process of joining materials to make objects from 3D model data, usually layer upon layer. It encompasses several synonymous terms like layered manufacturing and freeform fabrication. The AM process typically begins with a CAD model converted into a suitable file format (e.g., STL). This is followed by slicing, orientation adjustment, machine setup, and material deposition, culminating in post-processing and application. The shift from "What You See Is What You Get" (WYSIWYG) to "What You See Is What You Build" (WYSIWYB) is evident, emphasizing AM's digitally driven essence.

AM supports three primary applications: Rapid Prototyping (RP), Rapid Manufacturing (RM), and Rapid Tooling (RT). Its compatibility with low-volume, high-complexity production and just-in-time inventory strategies makes it a preferred choice across diverse sectors, including aerospace, healthcare, automotive, and education.

9.2.3 Intelligent Manufacturing: Evolution and Integration

The manufacturing trajectory has evolved through multiple industrial revolutions, with Industry 4.0 representing the latest shift marked by cyber-physical integration, IoT, and data-centric decision-making. IM encompasses the integration of real-time feedback loops, interconnected production units, and AI-enabled systems to optimize quality, efficiency, and adaptability.

IM covers the entire value chain, from conceptual design to end-of-life recycling, and is built on key technological pillars such as big data, cloud computing, AI, digital twins, and autonomous systems. It promotes system-wide integration, both vertically across different operational layers and horizontally across global value chains.

9.2.4 Integrating AM into Intelligent Manufacturing Systems

When integrated with IM systems, AM evolves from a standalone technology into a digitally orchestrated production tool. Sensors embedded in AM machines collect process data, which AI algorithms analyze in real time to predict defects, optimize build parameters, and reduce energy consumption. This closed-loop feedback enhances product quality and minimizes waste.

Moreover, cloud platforms and digital repositories enable global access to design files, facilitating distributed manufacturing. Digital twins simulate the build process virtually, enabling predictive maintenance, design validation, and reduced prototyping cycles.

Hybrid manufacturing chains, which combine AM with traditional manufacturing methods such as High-Speed Machining (HSM) or Electron Discharge Machining (EDM), further enhance surface finish and mechanical performance. These chains reduce lead time and cost while offering unparalleled design freedom.

9.2.5 Challenges in Advancing Intelligent AM

Despite its numerous advantages, integrating AM into IM ecosystems presents several critical challenges. One major issue is the lack of standardization across materials, data formats, and quality metrics, which limits interoperability and complicates certification processes, particularly in highly regulated industries. Another persistent challenge is achieving consistent material-process alignment, as maintaining uniform mechanical properties and microstructural integrity across different AM techniques remains difficult. Additionally, while real-time monitoring capabilities support in-process quality control, implementing high-resolution, multi-sensor fusion for adaptive control is still technologically complex. The integration of AM with enterprise-level systems such as ERP, MES, and PLM is further hindered by digital silos and fragmented software ecosystems. Moreover, the digital nature of AM introduces heightened cybersecurity risks, necessitating robust measures to protect sensitive design data and intellectual property. Finally, the successful deployment of intelligent AM systems depends on a highly skilled workforce with expertise in areas such as AI, robotics, and materials science, where talent shortages remain a significant bottleneck.

9.2.6 Future Outlook: Toward a Resilient and Responsive Manufacturing Paradigm

As the AM landscape matures, its integration with IM systems is poised to revolutionize the production ecosystem. Future factories will leverage AM for decentralized, on-demand, and sustainable production. This shift from Design for Manufacturing (DFM) to Manufacturing for Design (MFD) unlocks possibilities for shape, hierarchy, material, and functional complexity in design.

Collaborative innovation across academia, industry, and policymakers will also be essential to overcoming existing barriers. With advancements in CPS, AI, and data governance, AM will solidify its role as a key enabler of the smart manufacturing revolution, delivering customized, high-performance solutions while aligning with the goals of sustainability, agility, and operational excellence.

9.2.7 Strategic Pathways for Implementation

Organizations must adopt structured implementation strategies to ensure a successful transition to intelligent AM systems. This includes conducting a comprehensive assessment of current capabilities, identifying use cases where AM delivers maximum value, and aligning these with broader digital transformation goals. Pilot projects, especially in aerospace, healthcare, and tooling sectors, can serve as learning grounds before scaling.

Standardization efforts must also be supported by active participation in industry forums and collaborative research initiatives. Establishing benchmarks for process reliability, repeatability, and material properties will build trust and enable wider adoption. Furthermore, cloud-enabled design repositories and blockchain technologies can be leveraged for secure design sharing, traceability, and IP protection in globally distributed production networks.

9.2.8 Education, Training, and Workforce Development

A critical enabler for intelligent AM integration is the upskilling of the workforce. Academic institutions, vocational training centers, and industrial stakeholders must collaborate to design interdisciplinary curricula that span material science, digital manufacturing, robotics, and data analytics. Certifications and continuous professional development programs will be instrumental in equipping the workforce with the necessary competencies.

Simulation-based learning environments and AR-assisted training modules can enhance the learning curve, allowing students and workers to visualize AM processes, troubleshoot issues virtually, and understand the interdependencies between design, materials, and process settings.

This proactive approach to talent development ensures the long-term sustainability of IM ecosystems.

In a nutshell, the integration of AM with IM frameworks is more than a technological convergence; it represents a paradigm shift toward responsive, efficient, and sustainable industrial systems. While the path forward includes significant challenges, ranging from standardization and quality control to workforce readiness and cybersecurity, the benefits are transformative. Moreover, as Industry 4.0 evolves, AM's role within intelligent ecosystems will become increasingly vital. By embracing digital tools, fostering collaborative innovation, and investing in education and standardization, the vision of autonomous, intelligent, and sustainable manufacturing is well within reach. The synergy between AM and IM is not just reshaping how we make things; it is redefining what is possible in the future of manufacturing.

9.3 Circular Manufacturing

9.3.1 Introduction

In the context of Industry 4.0, the convergence of circular manufacturing and IM presents a transformative opportunity to reshape industrial systems for sustainability, efficiency, and resilience. Circular Manufacturing emphasizes minimizing resource input and waste output by promoting practices such as reuse, remanufacturing, recycling, and product life extension. IM, by contrast, leverages emerging digital technologies, such as CPS, AI, big data analytics, and the IoT, to create adaptive, self-optimizing production environments that respond in real time to changes in demand, resource availability, and operational conditions.

Together, circular manufacturing and IM form a synergistic framework that not only addresses the urgent need for sustainable production but also supports the strategic goals of competitive advantage, supply chain resilience, and regulatory compliance in an increasingly resource-constrained world. The intersection of these paradigms is not just technological; it is systemic, involving a fundamental rethink of how products are designed, produced, used, and recovered.

According to a recent bibliometric analysis,[2] global research interest in integrating circular economy principles into sustainable manufacturing is growing rapidly. With countries like India, Italy, and the United Kingdom leading in research output and platforms such as 'Sustainability Switzerland' and 'Procedia CIRP' publishing cutting-edge work, the integration of circular manufacturing and IM has emerged as a central focus for both academic research and industrial policy.

9.3.2 Integration Avenues

The integration of circular manufacturing and IM creates significant synergies by enabling closed-loop, data-informed material and energy cycles. IoT-enabled sensors embedded in products and machinery can track usage patterns and environmental conditions, enabling real-time insights into wear and tear or product end-of-life status. This data can then inform predictive maintenance schedules and guide decisions around remanufacturing or reuse.

AI and ML algorithms further support circularity by analyzing vast datasets to identify inefficiencies, forecast return flows, and optimize logistics for reverse supply chains. Digital twins allow manufacturers to simulate and optimize product designs for recyclability and modularity, reducing the need for physical prototypes and accelerating innovation cycles. Blockchain and cloud-based platforms enhance transparency and traceability across the circular value chain, enabling better stakeholder coordination and verification of material provenance.

Advanced reuse and repurposing strategies, such as those highlighted by Psarommatis et al. (2025)[3] are also critical components. Automated disassembly, sensor-embedded systems, and flexible supply chains facilitate product recovery and modular reassembly, while upcycling and product-as-a-service models encourage lifecycle extension. These enable real-time monitoring of returned products, ensure resource-efficient recovery, and minimize waste generation. Integrating digital product passports (DPPs) further enhances compliance with sustainability standards and supports efficient data exchange among manufacturers, regulators, and customers.

In addition to operational benefits, this integration drives circular business model innovation. For example, predictive analytics can inform rental-based models in which equipment maintenance and upgrades are pre-emptively scheduled, maximizing uptime and customer satisfaction. Service-oriented approaches, such as leasing and sharing platforms, reduce raw material consumption and incentivize durability and modularity in design. As IM platforms mature, they also enable dynamic pricing models and feedback loops to adjust product usage or recycling incentives in real time.

9.3.3 Challenges to Integration

Despite its promise, the integration of circular manufacturing and IM is met with several formidable challenges:

- Technological Readiness: Many manufacturers, especially SMEs, lack the digital infrastructure and technical capabilities required for intelligent circular operations. Capital investment also remains a barrier.
- Standardization and Interoperability: The absence of universal standards for data formats, product lifecycle metrics, and quality

benchmarks hinders seamless integration and regulatory approval, particularly in safety-critical sectors.

- Organizational Barriers: Resistance to change, departmental silos, and unclear ownership of circularity initiatives limit the scalability of integrated solutions.
- Economic and Core Acquisition Constraints: The variability in volume, timing, and quality of returned products makes core acquisition unpredictable.
- Policy and Regulation Misalignment: Regulatory frameworks are fragmented, with limited guidance for technologies spanning sustainability and smart systems. Global adoption of best practices, such as the European Union's (EU) End-of-life Vehicle (ELV) directive, remains uneven.
- Workforce Skill Gaps: The interdisciplinary nature of circular-intelligent systems demands expertise in materials science, data analytics, design for disassembly, and AI, a combination currently in short supply.

9.3.4 The Way Forward

To overcome these challenges, a multifaceted strategy is essential:

- Assessment and Benchmarking: Companies should begin by evaluating their digital and circular maturity using structured diagnostic tools, such as IMPULS Industry 4.0 Readiness Assessment [developed by the German Mechanical Engineering Industry Association (VDMA)], Digital Maturity Model (DMM) by Deloitte, Smart Industry Readiness Index (SIRI) by Singapore Economic Development Board, and BS 8001:2017 Maturity Matrix by British Standards Institution.
- Modular and Lifecycle-Oriented Design: Encouraging modularity in product design promotes ease of reuse, repair, and remanufacturing.
- Technology Implementation: Adoption of enablers such as AI-powered quality control, blockchain for traceability, and cloud-based data sharing should be aligned with clear business cases.
- Cross-Sector Collaboration: Universities, industries, and regulatory bodies must co-create standards and pilot frameworks that align with emerging circular-intelligent practices.

Education and training must evolve to include simulation-based learning, AR/VR-assisted disassembly modules, and interdisciplinary programs blending engineering, design, and sustainability. Governments should incentivize adoption through circular procurement policies, R&D funding, and certifications for reuse and repurposing systems. Companies should also leverage reverse logistics and design-for-recovery models to increase circular throughput and resilience against global supply chain disruptions.

Moreover, the creation of dedicated innovation hubs and testbeds for circular-IM could serve as practical platforms for piloting and scaling solutions. Lifecycle-based KPIs and sustainability reporting frameworks should be integrated into ERP systems to align decision-making with long-term value creation.

Conclusively, the fusion of circular manufacturing and IM represents more than a technological advancement. It signals a paradigm shift toward regenerative, adaptive, and forward-looking production systems. This integrated approach enables manufacturers to reduce waste, enhance resource productivity, and foster innovation across the value chain. Successful integration requires not only technology but also policy coherence, financial modeling, and robust implementation frameworks. Furthermore, organizations can dynamically respond to regulatory pressures and market demand by deploying digital tools to implement circular strategies. Moving forward, success will depend on collaborative innovation, workforce development, and regulatory realignment. The intelligent circular model thus offers a scalable, sustainable blueprint for building resilient manufacturing systems that meet the needs of both people and the planet.

9.4 Energy Efficiency

9.4.1 Introduction

As the manufacturing industry evolves in the era of Industry 4.0, energy efficiency has emerged as a central pillar in advancing sustainable, intelligent production systems. Smart factories, enabled by the integration of CPS, AI, big data analytics, and the IoT, are increasingly optimized for real-time decision-making, predictive maintenance, and resource-efficient operations.[4] These capabilities position IM as a catalyst for operational excellence and a foundational enabler of global sustainability efforts.

9.4.2 Current Trends in Energy Efficiency in Smart Manufacturing

Smart factories leverage interconnected technologies to achieve adaptive, self-regulating, and energy-efficient production. Key enablers include digital twins for simulation, edge and cloud computing for real-time processing, and IoT sensors for energy monitoring and load balancing. Large-scale foundation models (LSFMs) and ML techniques are increasingly being deployed to detect inefficiencies, forecast energy demand, and optimize asset utilization dynamically.[5]

Recent research emphasizes deploying advanced decision-making frameworks that incorporate MCDM methods, such as intuitionistic fuzzy sets

(IFS), to handle ambiguity and complexity in selecting energy-saving solutions. These systems allow manufacturers to align energy goals with production schedules, asset health, and environmental constraints.

Predictive analytics and peer-to-peer energy trading frameworks are also being explored to balance loads across multi-energy systems, reducing wastage and operational costs. Optimization techniques, such as methods based on removal effects (MEREC) and ranking of alternatives using functional mapping of criteria sub-intervals into a single interval (RAFSI), are increasingly employed to rank energy-efficient alternatives in a manner resilient to rank-flipping, which is a common issue in traditional MCDM systems.

9.4.3 Challenges in Achieving Energy-Efficient Intelligent Manufacturing

Despite significant progress, several barriers hinder the full realization of energy-efficient IM:

- Data Fragmentation and Noise:
  While IoT systems generate vast amounts of energy-related data, much of it is unstructured, noisy, and lacks labeling, posing difficulties for model training and accurate decision-making.
- Scalability and Generalization: Traditional deep learning models often perform sub-optimally when generalized across varied manufacturing settings. Their reliance on task-specific training limits cross-domain application.
- Integration Complexity: Many legacy systems are incompatible with modern AI and energy management modules, resulting in fragmented deployments and suboptimal performance.
- Ambiguity in Decision-Making: Criteria influencing energy decisions often conflict. For instance, maximizing throughput may counter energy-saving goals. Without robust frameworks such as IFS, managing these trade-offs remains difficult.
- Lack of Skilled Workforce: Integrating AI-based energy optimization systems demands expertise in manufacturing systems, AI modeling, and sustainability, a combination often scarce in traditional industries.

9.4.4 Way Forward and Best Practices

To advance energy efficiency in IM, a multipronged approach is required:

- Adoption of LSFMs: Foundation models, with their superior generalization and multi-modal capabilities, can process diverse data types and adapt to various factory conditions. Their application in predictive maintenance, fault detection, and energy demand forecasting can significantly improve efficiency.

- Data Governance and Fusion: Implementing standardized protocols for data collection, preprocessing, and integration ensures reliable inputs for AI models. Data lakes and federated learning architectures can help manage data across dispersed sources.
- Integrated Decision-Making Frameworks: Employing advanced MCDM methods such as RAFSI, MEREC, and intuitionistic fuzzy softmax Dubois-Prade (IFSDP) allows for robust, context-sensitive decision-making in energy optimization.
- Digital Twins and Simulation:These tools enable virtual testing of energy policies and retrofitting strategies before deployment, minimizin g risks and improving return on investment.
- Policy and Training Alignment: National policies must promote AI-driven energy efficiency through tax incentives, R&D grants, and skill development programs. Institutions should design hybrid curricula combining manufacturing engineering, AI, and sustainability.

9.4.5 Conclusion

Energy efficiency is no longer a supplementary goal in IM. It is a strategic imperative. The convergence of LSFMs, MCDM methods, IoT infrastructure, and sustainability frameworks paves the way for manufacturing systems that are not only smart but also green. Overcoming data integrity, integration, and workforce readiness challenges will require a collaborative effort between academia, industry, and policymakers. Nevertheless, the path forward is promising, as energy-efficient IM is critical in achieving sustainable industrial growth.

9.5 Reverse Engineering[6]

9.5.1 Introduction

Reverse Engineering (RE) is a structured methodology that deconstructs a physical product to obtain its design, material composition, and manufacturing process. It plays a pivotal role in replicating, improving, or understanding products, particularly when original documentation is unavailable or innovation demands adaptation from existing designs. In recent years, integrating IM technologies, such as AI, IoT, CPS, and big data analytics, has transformed the scope and effectiveness of RE methodologies.

9.5.2 Intelligent Manufacturing: A Catalyst for Reverse Engineering

The convergence of IM and RE has introduced a paradigm shift in how products are reverse engineered. Traditional RE relies heavily on manual scanning, CAD reconstruction, and model validation. With IM tools, this process becomes faster, more precise, and data-driven.

Smart sensors and IoT-enabled 3D scanners embedded in manufacturing environments facilitate high-resolution data acquisition in real time. These sensors capture geometry and record material behavior, surface conditions, and thermal characteristics, enhancing the fidelity of digital models. AI-powered algorithms analyze this multi-modal data to automatically clean, align, and convert point clouds into CAD-ready models, minimizing manual intervention in polygon modeling, NURBS surface generation, and solid modeling.

Digital twins, a hallmark of IM, play a key role in the RE workflow by simulating the scanned part's behavior under various operational conditions. This enables virtual validation, stress analysis, and iterative re-design without producing physical prototypes. Additionally, AI-based feature recognition tools help extract functional and geometric features such as holes, slots, and threads directly from the mesh data, thus accelerating design intent recovery.

9.5.3 Benefits and Current Practices

The incorporation of IM practices into RE enables:

- Accelerated Model Reconstruction: AI and machine learning reduce the time needed for surface reconstruction and feature detection.
- Improved Accuracy: Enhanced sensor integration minimizes errors during scanning and point processing, especially in complex or organic shapes.
- Knowledge-Based Re-Design: Using AI to learn from historical product failures, wear patterns, and re-design iterations supports intelligent product improvement.
- Remote and Cloud-Based Collaboration: Cloud platforms enable real-time sharing of 3D scans and models across global teams, streamlining collaborative RE and decision-making.
- Integration with CAD/CAE/CAM: Seamless data exchange between RE software and intelligent CAD/CAE environments allows faster prototyping and validation.

Current practices are increasingly leveraging hybrid models where machine learning algorithms suggest optimal scanning parameters based on material type and part geometry. AI-based decision systems are also used to determine whether a part should be repaired, replaced, or redesigned based on its structural health, economic feasibility, and functional relevance.

9.5.4 Challenges

Despite these advancements, several barriers remain:

- Data Overload and Noise: Intelligent RE systems generate large volumes of data that require effective filtering, fusion, and storage solutions.
- Software Compatibility: Integrating RE outputs with diverse CAD/CAM tools can be complex due to proprietary formats.
- Skill Gaps: The fusion of mechanical engineering, AI, and data analytics necessitates a workforce with interdisciplinary skills that are still developing in many institutions.
- Cost of Deployment: Intelligent RE systems require high initial investment in scanning equipment, edge computing, and AI software.

9.5.5 Way Forward

The way forward lies in embedding RE into broader IM ecosystems. Key strategies include:

- Standardization of Data Formats and Protocols to ensure interoperability among RE and IM tools.
- Adoption of Edge and Cloud Computing for real-time RE data processing and collaborative design.
- Integration of Predictive Analytics using historical data to anticipate component failure and reverse engineer upgrades proactively.
- Upskilling the Workforce offering interdisciplinary training that blends mechanical design, digital modeling, and AI fundamentals.

9.5.6 Conclusion

The integration of IM into RE significantly enhances the ability to replicate, adapt, and innovate from existing designs. By automating scanning, model reconstruction, and simulation processes, IM makes RE faster, more precise, and inherently more intelligent. This synergy is not only redefining traditional engineering workflows but also enabling agile product development in domains ranging from aerospace to biomedical engineering. As digital tools evolve, intelligent RE will become a cornerstone of sustainable and data-driven manufacturing.

9.6 Highlights

Highlights for the chapter on converging technologies for smart, circular, and energy-efficient manufacturing are as follows:

a. Computer Vision: Enables machines to interpret visual data using image acquisition, segmentation, 3D vision, and deep learning techniques. It is applied in real-time inspection, robotic guidance, and predictive maintenance within IM environments.
b. Additive Manufacturing (AM): Supports rapid prototyping and just-in-time manufacturing with digitally driven processes. Integration with IM systems enhances defect prediction, process optimization, and cloud-based design access.
c. Circular Manufacturing: Aligns with IM through IoT-based monitoring, digital twins, and AI-enabled reverse logistics, facilitating reuse, remanufacturing, and recycling within closed-loop production cycles.
d. Energy Efficiency: Achieved through deployment of LSFMs, MCDM-based decision-making, and digital twins for simulation-based optimization of energy-intensive processes.
e. Reverse Engineering (RE): Intelligent RE incorporates AI, IoT, and digital twins for efficient part digitization, feature recognition, and model reconstruction. Enhances innovation through knowledge reuse and functional adaptation.
f. Cross-Cutting Challenges: Include data interoperability, legacy system integration, skills gaps, and high deployment costs across all domains.
g. Strategic Enablers: Emphasize the need for standardization, cloud and edge computing, cross-sectoral collaboration, and interdisciplinary workforce training.
h. Future Outlook: Advocates for resilient, decentralized, and adaptive manufacturing systems, leveraging digital technologies to meet sustainability and operational excellence goals.

Figure 9.1 shows the converging technologies for intelligent manufacturing.

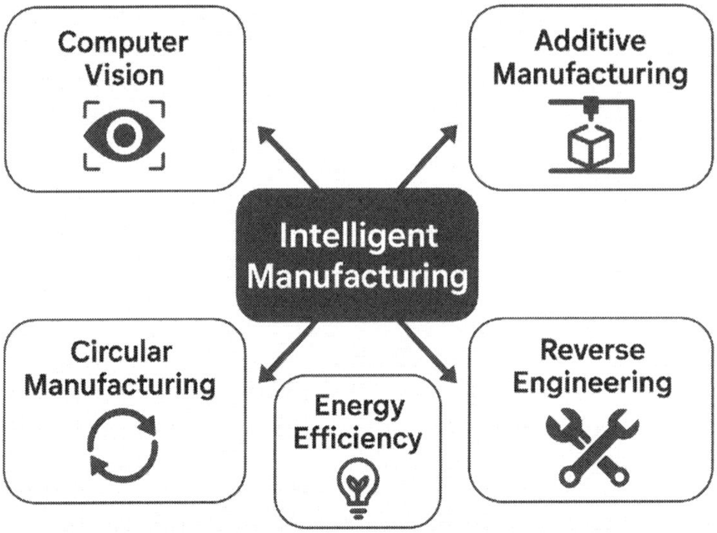

FIGURE 9.1
Converging technologies for intelligent manufacturing.

Notes

1 Szeliski, R. (2022). *Computer Vision: Algorithms and Applications*. Springer International Publishing. https://doi.org/10.1007/978-3-030-34372-9
2 Purwanto, M. H., Nugroho, R. A., Jati, H. S., & Wulandari, R. A. (2025). Circular economy and sustainable manufacturing: A bibliometric analysis. *IOP Conference Series: Earth and Environmental Science, 1441* (1), 012021.
3 Psarommatis, F., May, G., & Azamfirei, V. (2025). Product reuse and repurpose in circular manufacturing: A critical review of key challenges, shortcomings and future directions. *Journal of Remanufacturing*. https://doi.org/10.1007/s13243-025-00153-y
4 Anjum, M., Kraiem, N., Min, H., Daradkeh, Y. I., Dutta, A. K., & Shahab, S. (2025). Optimizing sustainability: Information aggregation in smart factories for energy efficiency enhancement. *Scientific Reports*, 15, 8603.
5 Zhang, H., Semujju, S. D., Wang, Z., Lv, X., Xu, K., Wu, L.,... & Ma, X. (2025). Large scale foundation models for intelligent manufacturing applications: A survey. *Journal of Intelligent Manufacturing*, https://doi.org/10.1007/s10845-024-02536-7
6 Aqeel, A. B., Aziz, M. I., Zaman, U. K., & Aafaq, N. (2023). Reverse engineering: Past, present, and future prospects. In Uzair Khaleeq uz Zaman, Ali Siadat, Aamer Ahmed Baqai, Kanwal Naveed, & Atal Anil Kumar (Eds.), *Handbook of Manufacturing Systems & Design: An Industry 4.0 Perspective*: 283–316, Florida: Taylor & Francis, ISBN: 9781032353210.

Index

Made in the USA
Las Vegas, NV
19 September 2026

30f5af65-1e35-4305-af87-fd245d24406aR01